UNDERSTANDING SPACE-TIME

This book presents the history of space-time physics, from Newton to Einstein, as a philosophical development reflecting our increasing understanding of the connections between ideas of space and time and our physical knowledge. It suggests that philosophy's greatest impact on physics has come about, less by the influence of philosophical hypotheses, than by the philosophical analysis of concepts of space, time, and motion and the roles they play in our assumptions about physical objects and physical measurements. This way of thinking leads to new interpretations of the work of Newton and Einstein and the connections between them. It also offers new ways of looking at old questions about a-priori knowledge, the physical interpretation of mathematics, and the nature of conceptual change. *Understanding Space-Time* will interest readers in philosophy, history and philosophy of science, and physics, as well as readers interested in the relations between physics and philosophy.

ROBERT DISALLE is Professor in the Department of Philosophy, University of Western Ontario. His publications include a contribution to *The Cambridge Companion to Newton* (2002).

UNDERSTANDING SPACE-TIME

The Philosophical Development of Physics from Newton to Einstein

ROBERT DISALLE

University of Western Ontario

CAMBRIDGE
UNIVERSITY PRESS

CAMBRIDGE UNIVERSITY PRESS
Cambridge, New York, Melbourne, Madrid, Cape Town, Singapore, São Paulo, Delhi

Cambridge University Press
The Edinburgh Building, Cambridge CB2 8RU, UK

Published in the United States of America by Cambridge University Press, New York

www.cambridge.org
Information on this title: www.cambridge.org/9780521857901

First published 2006
Third printing 2008
This digitally printed version 2008

A catalogue record for this publication is available from the British Library

ISBN 978-0-521-85790-1 hardback
ISBN 978-0-521-08317-1 paperback

For my parents, Richard DiSalle and Joan Malinowski DiSalle

*. . . L'amor del bene, scemo
del suo dover, quiritta si ristora*

Contents

Contents

Figures

Preface

This book concerns the philosophy of space and time, and its connection with the evolution of modern physics. As these are already the subjects of many excellent books and papers – the literature of the "absolute versus relational" debate – the production of yet another book may seem to require some excuse. I don't claim to defend a novel position in that controversy, or to defend one of the standard positions in a novel way. Still less do I pretend to offer a comprehensive survey of such positions and how they stand up in light of the latest developments in physics. My excuse is, rather, that I hope to address an entirely different set of philosophical problems. The problems I have in mind certainly have deep connections with the problems of absolute and relative space, time, and motion, and the roles that they play, or might play, in the history and future of physics. But they can't be glossed by the standard questions on space-time metaphysics: is motion absolute or relative? Are space and time substantival or relational? Rather, they are problems concerning how any knowledge of space, time, and motion – or spatio-temporal relations – is possible in the first place. How do we come to identify aspects of our physical knowledge as knowledge of space and time? How do we come to understand features of our experience as indicating spatio-temporal relations? How do the laws of physics reveal something to us about the nature of space and time?

I see two compelling reasons to focus on these questions. On the one hand, I believe it will give us a more illuminating picture of the connection between the metaphysics of space and time and the development of philosophy in general. Historically, there have been two significant attempts to integrate the physics and the philosophy of space and time with a general theory of knowledge: Kant's critical philosophy, in its attempt to comprehend Euclid and Newton within a theory of the synthetic a priori; and logical positivism, in its attempt to comprehend Einstein within a conventionalist view. These attempts are widely recognized as failures, and I don't intend to try to rehabilitate them. But I believe that there is some insight to be gained from a better understanding of why they failed; more important, I hope to show that the task in which they failed – to explain the peculiar character of theories of space and time, and the peculiar role that they have played as presuppositions for the empirical theories of physics – is no less important for us than it was for them, and, moreover, is more nearly within our grasp. On the other hand, I believe that focusing on these questions will give us a clearer picture of the history of physics. For, as I hope to show in the following chapters, the moments when such questions have become most urgent are precisely the most revolutionary moments in the history of space-time physics. The great conceptual transformations brought about by Newton, Einstein, and their fellows simply could not have happened as they did without profound reflection on these very questions. And our sense that these transformations were crucial steps forward – that, apart from increasingly useful theories, they actually yielded *deeper understanding* of the nature and structure of space-time – has everything to do with the success of their philosophical work.

This is not an entirely novel idea. Something like it was at the heart of the positivists' interpretation of relativity theory: Einstein introduced special and general relativity by some "philosophical analysis" of the concepts of space and time. But this interpretation was based on a rather simplistic picture of relativity, as well as simplistic notions of what a "philosophical analysis" could be. Given the inadequacies of the positivists' attempt to put relativity into philosophical perspective, it has since appeared easier to see the relevance of philosophy to physics in simpler terms: as a source of philosophical *motivations* for physicists, and even of theoretical *hypotheses*, but not as a method of scientific *analysis*. For such motivations and hypotheses, it would seem, are inescapably subjective, and their objective worth can only be judged by the empirical success of the theories that they produce. Einstein thought that anyone who followed the philosophical steps that he had taken, whatever their scientific background, would be convinced of the basic principles of special and general relativity. By the later twentieth century, however, philosophers came to think of those steps as somewhat arbitrary, and as not very clearly related to the theories that Einstein actually produced. They had a heuristic value for Einstein, and may have again for a future theory of space-time. To believe again that such philosophical arguments could be crucial – not only to the motivation for a theory, but also to its real significance in our scientific understanding of the world – we need a more philosophically subtle and historically realistic account of those arguments, and the peculiar roles that philosophy and physics have played in them.

That is what this book aims to provide. It is not distinguished by any technical arguments or results; it benefits, in that regard, from the tradition of important works on absolute and relational space-time, such as Sklar (1977), Friedman (1983), and Earman (1989), that have done so much to make space-time geometry a familiar part of philosophical discourse. Nor can it claim to offer a wealth of previously unknown historical detail, although it does emphasize some historical figures who are rarely considered in the philosophy of space and time. Instead, this book seeks to present some fairly familiar developments from a completely unfamiliar perspective, as part of a remarkably concerted and coherent philosophical effort – an effort to analyze, from a series of critical philosophical standpoints, the evolving relationship between our physical assumptions and our knowledge of space and time. Early twentieth-century philosophers had a difficult time seeing the history from this perspective, because they saw the philosophy of space and time as essentially an argument against Newton, that is, as a struggle of modern epistemology against old-fashioned metaphysics. What this book attempts to show is that the best philosophy of space and time – the part that has been decisive in the evolution of physics – has been a connected series of arguments that *began* with Newton, arguments about how physics must *define* its conceptions of space and time in empirical terms. By viewing the history in this way, my book proposes to shed some light on other questions that were puzzling to twentieth-century philosophy of science: above all, how the transformation of fundamental concepts, like those of space, time, and motion, can be understood as a rational development.

The most obvious audience for this book, then, would be philosophers of science with an interest in physics, and physicists with an interest in the conceptual development and the philosophical significance of their discipline. But I hope that it will also be of interest to any philosophical reader who is curious about the role of philosophical analysis as a tool of scientific inquiry, and about the physical world as an object of philosophical reflection. On both of these matters, the history of the physics of space and time is an unparalleled source of insight.

Many people helped me with the writing of this book, but none more than William Demopoulos, my colleague and friend for nearly two decades. This book is indebted, not only to my many discussions with him and his careful reading of every draft, but also to the influence and the model of his own work on the foundations of mathematics and science.

I also owe a great debt to Michael Friedman, partly because of his constant guidance and encouragement of this project, but mostly because, like much of my work, it has deep roots in my philosophical engagement with his.

I particularly thank Howard Stein and David Malament, who supervised my graduate studies long ago, and who tried to teach me, by word and example, what the history and philosophy of science might aspire to. I hope that they will be able to discern something of their influence in my work.

It was Laurens Gunnarsen who, with his extraordinary gifts of mathematical intuition and patience, guided my very first steps on the path that eventually led to this book, and imparted to me his love of its subject matter.

Others who contributed to this book at some stage or other, directly or indirectly, include: John Bell, Martin Carrier, Darcy Cutler, Mauro Dorato, Michael Hallett, Ulrich Majer, Paulo Parrini, Miklos Redei, Heinz-Jürgen Schmidt, George Smith, and Gereon Wolters. My thanks to all of them, and to everyone who patiently heard my talks at various colloquia over the past several years, as the ideas for this book were evolving.

I would also like to thank Hilary Gaskin, of Cambridge University Press, for her early interest in this project and encouragement of it. I thank the anonymous referees for the Press for their helpful suggestions, and Sarah Lewis and Anna-Marie Lovett for their editorial efforts. I am also indebted to Sona Ghosh for her thoughtful and intelligent work on the index.

Most of this book was written while I was a Senior Fellow at the Dibner Institute for the History of Science and Technology. I will always be grateful to the staff of the Institute, the other Fellows from 2002–2003, and especially the Acting Director, (again) George Smith, for creating an ideal intellectual atmosphere in which to pursue this project. I also thank the Dibner family for their tradition of support for work of this kind. Additional financial support came from the Social Sciences and Humanities Research Council of Canada.

I would like to thank my son Christopher and my daughter Sofia, for giving me the best reasons to undertake this work and all the right encouragement to finish it.

Last, and most of all, I would like to thank my wife Zanita. She lived with this project and supported it from its earliest beginnings; what she has given to this book, and to me, there will never be space or time enough to say.

CHAPTER I

Introduction

Why is there a "philosophy of space and time"? It seems obvious that any serious study of the nature of space and time, and of our knowledge of them, must raise questions of metaphysics and epistemology. It also seems obvious that we should expect to gain some insight into those questions from physics, which does take the structure of space and time, both on small and on cosmic scales, as an essential part of its domain. But this has not always seemed so obvious. That physics has an illuminating, even authoritative, perspective on these matters was not automatically conceded by philosophy, as if in recognition of some inherent right. No more did physics simply acquire that authority as a result of its undoubted empirical success. Rather, the authority came to physics because physicists – over several centuries, in concert with mathematicians and philosophers – engaged in a profound philosophical project: to understand how concepts of space and time function in physics, and how these concepts are connected with ordinary spatial and temporal measurement. Indeed, the empirical success of physics was itself made possible, in some part, by the achievements of that philosophical effort, in defining spatio-temporal concepts in empirically meaningful ways, often in defiance of the prevailing philosophical understanding of those concepts. In other words, the physics of space and time has not earned its place in philosophy by suggesting empirical answers to standing philosophical questions about space and time. Instead, it has succeeded in redefining the questions themselves in its own empirical terms. The struggle to articulate these definitions, and to re-assess and revise them in the face of changing empirical circumstances, is the history of the philosophy of space and time from Newton to Einstein.

That history is not usually understood in these terms. More commonly, it is identified with the history of the "absolute versus relational" question: are space, time, and motion "absolute" entities that exist in their own right, or are they merely abstracted from observable relations? Without doubt this has been an important question, both for physics and for philosophy, and philosophical stances on it have evidently been powerful motivating principles for physical speculation. For that reason it plays a large role in the history that I have to tell. But it is not the entire story, or even the central part. And the tendency to see the history of space-time theories through the lens of this controversy – a tendency that has prevailed for most of the past century or more – has therefore clouded our view of that history. The absolute–relational debate is a cherished example of the influence of philosophy on the evolution of physics, for it seems to exhibit fundamental theoretical physics in the aspect of a kind of inductive metaphysics, in which physical arguments are brought in support of metaphysical ideas, and vice versa, in an ongoing philosophical dialectic. But the struggle to define a genuine physics of space and time has involved another sort of dialectic altogether: not between metaphysical positions, but between our theory of space and time, as expressed in the laws of physics, and our evolving knowledge of matter and forces in space and time. The revolutionary changes in conceptions of space and time, such as those brought about by Newton and Einstein, were therefore driven by a kind of conceptual analysis: an analysis of what physics presupposes about space and time, and of how these presuppositions must confront the changes in our empirical knowledge and practice.

By overlooking this process of conceptual analysis, we tend to misrepresent the historical discussions of space and time by Newton, Einstein, and others, and the philosophical arguments that they gave; we fail to get a proper sense of the progressive force of those arguments, as central aspects of the scientific argument for theoretical change in the face of empirical discovery. But we do not merely cloud the historical picture. We also obscure the connections between the problems of space and time and some broader issues in the history of philosophy: the nature and function of a-priori presuppositions in science, and the rational motivations for conceptual change in science. To clear away these obscurities is the purpose of my book.

The revival of metaphysical debate on space and time, over the past several decades, must be understood as part of the general reaction against logical positivism in the late twentieth century. The positivist view was that debate had been largely settled by Einstein: clear-sighted philosophers had always grasped the relativity of space, time, and motion on epistemological grounds, and Einstein finally brought their insight to fruition in a physical theory. From the more recent literature on the absolute–relational controversy, by contrast, we get a more vivid and realistic picture of the interaction between physics and philosophy, especially of the diverse ways in which purely philosophical convictions have motivated some of the most revolutionary work in physics. And we see, moreover, how sometimes the philosophical aims of physicists have been unrealized – how much divergence there has been between the original philosophical motivations behind revolutionary theories, and the content and structure of the theories that were eventually produced. The most familiar example – and the most damning to the positivists' neat picture – is the divergence between Einstein's vision of a theory of "the relativity of all motion" and general relativity itself, which turned out to have similarities with Newton's theory of absolute space that Einstein found philosophically hard to accept. In such cases there can be no doubt of the tremendous heuristic power of the original philosophical ideas, yet they can give rise to theories that seem to contradict them.

This seemingly mysterious circumstance has a broader significance for the philosophy of science. A primary preoccupation of the philosophy of science, since the later twentieth century, has been the question of the rationality of scientific revolutions, and the commensurability or incommensurability of competing conceptual frameworks, a kind of question raised most forcefully by Kuhn (1970a). As a matter of the history and sociology of science, it is beyond dispute that there have been, and are, competing groups within scientific disciplines with competing aims and methods, and with finite capacities for communication and mutual understanding. As a matter of philosophy, however, Kuhn introduced the radical claim that scientific conceptual frameworks are by their very nature incommensurable with one another. Whatever one thinks of Kuhn's view, it should be clear that theories of space and time provided Kuhn with some of the most vivid examples of profound conceptual shifts – not merely dramatic shifts in beliefs about the world or even in scientific methods, but in the very concepts that define the objects of scientific inquiry, the phenomena to be observed and the magnitudes to be measured. Kuhn emphasized the transition from Newtonian to relativistic mechanics, for example, less because it challenged specific traditional beliefs than because it created a conceptual system within which fundamental concepts of length and time, and with them force, mass, and acceleration, would have to be revised (Kuhn, 1970a, p. 102).

This last notion was hardly original with Kuhn. On the contrary, it was a central point – one might even say, *the* most fundamental motivating principle – for the logical positivists' interpretation of Einstein. If special relativity had appeared to be a merely incremental change from Newtonian mechanics (or general relativity from special relativity), part of a gradual and cumulative development driven by the steady application of traditional scientific methods, it would have seemed to them completely without philosophical interest. It was precisely because Einstein had undertaken a radical revision of fundamental concepts that the logical positivists saw him as revolutionary for philosophy as well as for science. What distinguished Kuhn from the logical positivists, especially, was his view of how and why such conceptual revisions take place. According to Kuhn, "critical discourse" about the foundations of theories typically takes place because the prevailing theoretical framework is in crisis: from one side, it faces an accumulation of anomalies, or "puzzles" that ought to yield to the framework's standard methods, but that have somehow resisted being solved; from the other side, it faces serious competition from a novel alternative framework. "It is particularly in times of acknowledged crisis," Kuhn wrote, "that scientists have turned to philosophical analysis as a device for unlocking the riddles of their field," even though they "have not generally needed or wanted to be philosophers" (Kuhn, 1970a, p. 88). It was "no accident," therefore, that the twentieth-century revolutions against Newtonian physics, and indeed Newton's own conceptual revolution, were "both preceded and accompanied by fundamental philosophical analyses of the contemporary research tradition" (Kuhn, 1970a, p. 88). While he acknowledged the creative influence of philosophical analyses, however, Kuhn was not prepared to admit that a philosophical argument against an existing theory could furnish any objective argument on behalf of a new rival. Nor could he acknowledge that such arguments could illuminate the relations between the theories, or the sense in which the shift from the old to the new theory might be understood as genuine theoretical progress. Philosophical beliefs, in short, functioned in scientific revolutions as subjective influences; they might motivate or persuade individual scientists – making particular theories or lines of research more psychologically accessible or appealing for scientists of particular philosophical tastes – but could never provide anything resembling a rational justification for theory change. For the philosophical arguments for a particular paradigm are always based on the paradigm itself. "When paradigms enter, as they must, into a debate about paradigm choice, their role is necessarily circular. Each group uses its own paradigm to argue in that paradigm's defense . . . Yet, whatever its force, the status of the circular argument is only that of persuasion" (Kuhn, 1970a, p. 94). When scientists at a time of crisis "behave like philosophers," in Kuhn's phrase (1970b, p. 6), this is because they are engaging in inconclusive "debates about fundamentals" such as are characteristic of philosophy (Kuhn, 1970b, p. 6). The prominence of philosophical considerations during revolutionary times merely highlights the lack of any clear methodological rules to guide conceptual change.

For the positivists, by contrast, such a revision could have an objective *philosophical* ground, as a radical critique of concepts that were epistemologically ill-founded.[1] For example, relativity theory was motivated by, and embodied, an evident progress in the philosophical understanding of space and time and the ways in which we measure them. The revised concepts of mass, length, and time were not merely the side-effects of a change in world view, but, rather, direct expressions of this improved understanding. So the theory was not only motivated, but also justified, by the philosophical arguments of Einstein. There could be no question of the rationality of a conceptual transformation that appeared so clearly to be a kind of conceptual *reform*.

From the perspective of the later twentieth century, however, this understanding of Einstein's revolution seemed particularly misguided. On the one hand, it seemed to exemplify what was wrong with the positivists' approach to science in general: the simple-minded belief that unobservable theoretical entities could be eliminated, and that theory could be reduced to its purely empirical content. On the other hand, it exhibited mistaken views about the content of general relativity itself. A number of physicists and philosophers quickly noted this discrepancy, and appreciated the important continuities between general relativity and its predecessors. But the dominant voices in the emerging discipline of "philosophy of science" were those of the positivists, especially Reichenbach (1957); as a result, a proper understanding of the bearing of general relativity on the metaphysics of space, time, and motion was slow in coming. By the late 1960s, the elements of a more circumspect viewpoint were in place: that Newton's theory of absolute space and time was not a mere metaphysical appendage to his physics, but had some genuine foundation in the laws of motion; that general relativity did not "relativize" all motion, but distinguished among states of motion in radically new ways; and that space-time in general relativity was in some respects the same sort of metaphysical entity as it had been in Newtonian mechanics – at the very least, both theories characterize space-time geometry as an objective physical structure. In short, Einstein's work no longer seemed to have settled the absolute–relational controversy decisively in favor of relationalism. Therefore it no longer seemed to conform to the positivists' picture of it, as an epistemological critique that eliminated metaphysics from physics; that picture had only displayed their flawed understanding of the theory, and of the role of theoretical entities in science.

If general relativity is separated in this manner from the original philosophical arguments for it, then the arguments are relegated to the status
of mere subjective factors in the development and the acceptance of the
theory. From the point of view of the absolute–relational debate, this is
not a disagreeable outcome. It suggests that a theory of space and time is,
after all, a theory like any other, and that scientists will develop or accept
such a theory for the same kinds of reason as they would any other theory.
The metaphysical questions about space and time may then be translated
into a straightforward form: what does our best current physics say about
space and time? Rightly rejecting the positivists' view of relationalism as
the inevitable result of progress in epistemology, contemporary literature
views it (and its antithesis) essentially as a metaphysical *hypothesis*, confirmed or not by how well it accords with the best available physical theory.
This new attitude clearly implies that it is not for "the philosophy of space
and time" to judge what might be the best available theory. Physics presumably has empirical methods for deciding such things, and these are of
the highest philosophical interest – from them, if from anywhere, must
come the answer to Kuhnian concerns about incommensurability – but
the philosophical discussion of space and time may take such decisions for
granted. It is also implied, therefore, that what makes a theory "the best"
has nothing to do with its philosophical implications concerning space
and time. Philosophical "intuitions" might move physicists to prefer one
metaphysical hypothesis to another, and to try (as Einstein did) to create a
theory that accords with it, but the theory itself would have to be judged
on largely empirical grounds. An abstract philosophical argument against
"absolute" structures has no force; what relationalism needs is a theory that
can save the phenomena without them.

This, at any rate, is the implicit philosophical principle of the most
prominent recent literature. (For a contrasting view to which my own view
is indebted, see Friedman, 2002b.) In explicit form it can be traced back
at least as far as Euler, who, indeed, expressed it as clearly as anyone. We
don't possess, he argued, any principles of metaphysics that we can claim
to know as securely as we know the laws of physics (see Euler, 1748).
Therefore no metaphysical principles can possibly claim the authority to
question the laws of physics; in particular, a conception of space or time
that has a foundation in the laws of motion is inherently secure against
criticisms from metaphysical grounds, which are necessarily less secure and
more controversial than the laws of motion. Euler's specific target was Leibniz's objections to Newton's theory of absolute motion, a theory which, as
Euler clearly recognized, rested on physical laws that were considerably

better founded than anything in Leibnizian metaphysics. But his point was quite general, and from the point of view of our own contemporary literature, even too obvious to require any mention. The consensus appears to be that general metaphysical and epistemological arguments for absolutism or relationalism are of secondary interest, useful for historical and heuristic purposes. In place of such considerations, there is a general metaphysical assumption that real entities are just those postulated by the best current physics, and an epistemological assumption that just those ontological distinctions are meaningful that the best current physics is capable of making. (The discussions of Einstein's "hole argument" in the recent past exhibit these assumptions especially clearly, see Earman, 1989.) So the debate between relationalism and absolutism (or substantivalism) effectively reduces to the question, which of these positions is best supported by current physics? Answering this question involves great technical and conceptual challenges, but the question has become, in a *philosophical* sense, relatively straightforward.

There can be no doubt that this change is largely for the better. That discussions of space and time are ultimately accountable to the physics of space and time is probably beyond dispute, and is in any case (as I hope will be clear to the reader) a principle that this book shares with most of the philosophy of physics literature. I do suggest, however, that in the application of this principle, the role and significance of philosophical analysis has been overlooked. And this has created at least three interconnected problems. First, and most obviously, it encourages a distorted view of the actual history: instead of seeing the actual philosophical arguments of Newton and Leibniz in their original context, we forcibly translate them into terms that will allow us to compare them against current physics. One might ask, of course, do such positions have any present philosophical interest if they *cannot* be translated into something relevant to contemporary physics? Conversely, if what we now call "absolutism," substantivalism," and "relationalism" are of demonstrable relevance to current physics, does it really matter whether they have any genuine connection with the sides of an ancient debate? Frankly, it is not the primary purpose of this book – though it is an indispensable part of my task – to defend historically accurate interpretations of Newton, Leibniz, and their fellows, and to distinguish their views carefully from the modern positions. The misinterpretations are important only because they have distracted our attention from the most important problems that Newton and Leibniz – along with Kant, Mach, Poincaré, Einstein, and others – were trying to address. These problems concerned, not whether space and time are absolute, but *how questions about*

space and time are to be framed in the first place. How is objective knowledge
of spatial and temporal relations – let alone of "space itself" or "time itself" –
possible? What does it mean to attribute some particular structure to space
or time? What is the status of the basic principles of geometry – how does
axiomatic geometry become an empirical science? How do concepts such
as absolute space and absolute time acquire some empirical meaning?

Second, by overlooking these questions, we overlook the relevance of
theories of space and time to a broader philosophical question: the nature
and status of a-priori knowledge. The relevant issue is not, as one might
suspect, whether we have some knowledge of space and time that is prior
to all experience. Rather, it is whether, and how, theories of space and time
have functioned as conceptual frameworks, that is, as formal structures
that *define* physical properties as empirically measurable magnitudes. If
theories of space and time thus function as presuppositions for empirical
inquiry, then the arguments for the theories themselves must be something
other than empirical arguments of the familiar inductive or hypothetico-
deductive sort. In the post-positivist era, it is common to see all theories –
even, for some philosophers, mathematics and logic as well as fundamental
physics – as forming a "man-made fabric which impinges upon experience
only along the edges" (Quine, 1953, p. 42). This suggests that there is only
a difference of degree between abstract theoretical principles and statements
of empirical fact; when "a conflict at the periphery occasions readjustments
in the interior of the field," there is no principled way to decide which
beliefs ought to be revised. It follows that every principle within the fabric
is to some extent an empirical hypothesis. Whatever the merits of this view,
it hardly helps us to understand the conceptual development of theories
of space and time. For those whose work had the greatest impact on that
development – from Newton and Kant to Poincaré and Einstein – certainly
were convinced that concepts of space and time had a special status, as the
presuppositions required for an intelligible account of matter and forces.
They believed, therefore, that their revolutionary work required explicit
reflection on how the concepts of space, time, and motion must be defined,
in order that questions about the nature of matter and force might become
empirical questions.

This leads us to the third problem, which is the problem of conceptual change. If the revolutionary developments in the theory of space and
time involved changes in the meanings of fundamental concepts, then it
will be difficult to meet the challenge posed by Kuhn, and to show that the
acceptance of new theories is a rational scientific choice. Obviously the consequences of Newtonian mechanics, for instance, can be tested empirically

with great precision. In order even to formulate those consequences for any real system, however, we first have to accept a series of *interpretive* principles: for example, that every acceleration is to be regarded as a measure of the action of some force. While this principle makes possible the empirical analysis of motion, it cannot be the object of such an analysis itself; we cannot perform tests to see whether forces conform to the principle, for it is a criterion by which we identify force in the first place. This is what led Poincaré to characterize the laws of motion as "definitions in disguise": they appear to make empirical claims about the nature of force, but in fact we cannot say what a force is except by stating the laws. The interpretive character of such principles, in fact, is the key to their role as a-priori presuppositions. But this raises the question how the introduction of such principles or, even more, a radical change in them, can be justified on any scientific grounds. For the logical positivists, interpretive principles were a matter of conventional choice: a physical theory is a purely formal mathematical structure, and to interpret it is to make some arbitrary stipulation about how its formal elements are to be "coordinated" with observation (see Carnap, 1995). In the case of space-time geometry, the role of stipulations was supposed to be particularly central. For, within a given geometrical framework, physical magnitudes can be measured empirically, but the framework itself is not fixed until we agree on the meaning of geometrical magnitudes such as length and time. If this view has few followers now, it should be remembered that, in the early twentieth century, it seemed to have the support of Einstein himself, who sometimes suggested that special relativity rested on an arbitrary stipulation about time. Einstein's great conceptual transformation, on this view, replaced the ill-defined concepts of Newtonian physics by unequivocal "coordinative definitions" of simultaneity, length, and time.

If they are arbitrary, however, these stipulations can only be judged by the success of the framework that they help to define. As Carnap would put it, such a framework defines a set of "internal questions" and a set of objective criteria for answering them; whether to adopt or abandon any given framework is an "external" question that can only be answered on pragmatic grounds such as overall simplicity and utility (see Carnap, 1956). By those criteria, it would be hard to deny that Newton's theory or Einstein's, in the long run, turned out to be better than what it replaced. But that sort of judgment is not necessarily straightforward, or even possible, at the time of a theory's acceptance; sometimes it is only made possible by the sustained efforts of those who have accepted the theory from the outset. This is the kind of historical situation that Kuhn portrayed so convincingly:

whatever their ultimate success, theories are often accepted while they are in a somewhat inchoate state, by scientists who have faith but little evidence that they will succeed in the long run. It would appear, in short, that no simple methodological rule can justify the decision to interpret the world in a novel way, even though the benefits of doing so might eventually become obvious. The logical positivists never faced this difficulty, because, again, they viewed special and general relativity as inherently progressive, having finally connected physics with modern insights into the epistemology of geometry. But in the aftermath of Kuhn, when the positivists' philosophical case for relativity is seen as a mere subjective preference – at most, a useful heuristic principle rather than a rational ground – the difficulty arises again.

Post-positivist philosophy of science does not take problems of interpretation very seriously, because of its rejection of the positivists' theory of theories. Instead of seeing a scientific theory as a set of axioms, which depend on coordinative definitions (or "correspondence rules," "meaning-postulates," etc.) for their connection with experience, contemporary philosophy of science typically represents a theory as a model-theoretic structure. Reference to experience is expressed in the hypothesis that the theory, understood as a structure, has "the world" as one of its models. As a way of talking about theories, this "semantic view" has definite merits, some of which will be discussed (and occasionally exploited) later in this book. But it is not a way of understanding the physical interpretation of formal structures; on the contrary, it tends to hide the problem from view.[2] By asserting that "the world is a model of Euclidean geometry," for example, we are simply taking for granted what the positivists saw the need to define: what does it mean for the world to be a model of Euclidean or any other geometry? What in the world is a straight line or an invariable length? How, in general, are we to decide which observable objects are to stand for which geometrical structures?

It should be clear, now, why questions like these have a profound bearing on the three problems I named. They are crucial to understanding the development of space-time theory, because the most important and historic philosophical arguments about space, time, and motion – those that have had the greatest impact on the evolution of physics – have arisen precisely at times when these questions have become urgent. Addressing them, in such circumstances, has engaged physics in a profound examination of its own a-priori presuppositions. In such circumstances, the emerging philosophical arguments have been, more than mere defenses of metaphysical preference, agents of conceptual transformation. The problem is to understand how arguments about the definitions of spatio-temporal concepts – about the principles that constitute for us the very objects of scientific study – can possibly be objective scientific arguments, and how the resulting transformation can be understood as a deeper insight into the nature of space and time.

To solve this problem was, arguably, a central aim of Kant's critical philosophy, and the fate of his attempt is particularly instructive. In his view, the argument for a fundamental constitutive principle was a transcendental argument, showing that the principle is a condition of the possibility of experience; the argument for the Newtonian framework of space and time, accordingly, was that it was the condition for our understanding of matter, motion, and force. So Newton's revolution was justified by the fact that it articulated, for the first time in the history of science, concepts of space, time, and causality by which the entire Universe could be understood as an interacting system. Traditional metaphysics, and even common sense, by contrast, stood revealed as having only the most confused ideas of these matters – except to the extent that something like the Newtonian conceptions were latent in them. Like the positivists' interpretation of Einstein, however, this interpretation of Newton now seems to epitomize the shortcomings of the philosophy that produced it. Kant understood rightly that the Newtonian principles, as presuppositions of empirical reasoning, could not themselves be derived by empirical reasoning of the same kind. But he mistakenly inferred that they are immune from any empirical reasoning – that they are connected with the fundamental categories of human understanding, and hence are both necessary and sufficient for *any* intelligible account of our experience. As the later career of Newtonian physics suggests, constitutive principles can be overturned by empirical knowledge. They cannot be fixed for all time, any more than they can be changed arbitrarily.

The example of Kant gives us, at least, some idea of what is at stake here. If we could reconcile the apparently conflicting aspects of the principles of space-time geometry – that they are constitutive and interpretive, yet somehow contingent upon our evolving empirical knowledge – we would be on the way to understanding theoretical interpretation as a rational scientific process, and radical change of interpretation as (at least sometimes) a kind of scientific progress. Furthermore, beyond these problems in the philosophy of science, we would gain some insight into a general philosophical question to which neither Kant nor the positivists had a convincing answer: why must the metaphysics of space and time answer to physics at all? What gives physics its authority in these matters? Fortunately, the keys to answering these questions can be found in the history of physics. To understand how the conceptions of space and time have been defined and redefined, in the emergence of modern physics, we need to re-examine the arguments by which those definitions were introduced by people such as Newton and Einstein. The definitions arise, not from arbitrary stipulation, but from conceptual analysis – from a dialectical engagement with existing ideas of space and time, revealing their hidden presuppositions and confronting them with new observation and theory. The radical changes in meaning are not, as Kuhn suggested, mere side-effects of theory change; they are the results of deliberate and self-conscious philosophical analysis that are themselves the engines of theory change. And their impact on philosophy – their authority to challenge existing philosophical notions of space and time – comes from the fact that they confront those notions on philosophical grounds, and expose their implicit connections with assumptions about physics.

That, at least, is the history I will present of the theories of space, time, and motion since Newton. It is not, therefore, another retelling of the story of the absolute–relational controversy. Rather, it is an account of how concepts of absolute and relative space, time, and motion have come to play the parts that they play in physical theory, and the impact that the construction, refinement, and critical analysis of these concepts has had on the conceptual development of physics. It is therefore no less than the story of the movement of physics toward a kind of philosophical maturity – toward a state of clarity in fundamental concepts, and of self-consciousness concerning the ways in which fundamental concepts acquire their empirical meaning.

NOTES

1. For a similar but more persuasive account of the role of conceptual criticism, to which my own account is indebted, see Torretti (1989, section 2.5).
2. My discussion follows that of Demopoulos (2003).

CHAPTER 2

Absolute motion and the emergence of classical mechanics

At a time when the relativity of motion was just beginning to be understood, Newton introduced a theory of absolute motion in absolute space and time. The controversy that then began has never ceased. What right did Newton have to explain the observable relative motions by an appeal to these unobservable entities? What role can such metaphysical hypotheses play in empirical science? By re-examining Newton's arguments for his theory, and understanding its role in the science that he helped to develop, we can see that these questions are misdirected. Newton's theory of space and time was never a mere metaphysical hypothesis. Instead, it was his attempt to define the concepts presupposed by the new mechanical science – the conceptual framework that made relative motion physically intelligible within a conception of causal interaction. Rather than an empirically questionable addition to his scientific work, it was an essential part of his work to construct an empirical science of motion. Rather than mere metaphysical baggage carried by an otherwise empirically successful theory, it was inseparable from Newton's effort to define the empirically measurable quantities of classical mechanics.

2.1 NEWTON AND THE HISTORY OF THE PHILOSOPHY OF SCIENCE

The history of Newton's ideas of space and time was once part of a philosophical justification for general relativity. For much of the twentieth century, the standard view of that history was something like this. When Newton introduced the theory, it was immediately obvious to his wisest philosophical contemporaries that this was a backward step. The Aristotelian conception of the universe as a collection of distinguished places, to which bodies belonged according to their particular qualities, had given way to the conception of an infinite, homogeneous Euclidean space; the conception of types of natural motions, all defined in relation to the resting

13

Earth, had given way to the recognition that motion is essentially relative, i.e. is nothing more than the relative displacement of a body relative to other bodies. The second point had been absolutely essential, in fact, to overcoming what had otherwise seemed to be good empirical arguments against any motion of the Earth. Therefore Newton's ideas of absolute space and absolute motion represented just the sort of primitive meta-physical thinking – a kind of reification of abstract objects – from which physics was now trying to escape, in order to become an empirical science. Huygens and Leibniz were particularly emphatic in rejecting these ideas. But Newton, through his notorious "water-bucket" experiment, claimed to know how to determine true motion *dynamically*: the centrifugal forces that arise in the spinning bucket demonstrate that the water is rotating, not merely relative to its material surroundings (the local frame of reference), but with respect to space itself. Leibniz and Huygens, along with a few other philosophers such as Berkeley, could easily see the emptiness of such an argument, which invoked a mysterious unobservable entity to explain the observed phenomenon. And they could see the inherent inability of physics to say anything meaningful about motion without referring it to observable objects. What they could not see was how to construct a dynam-ical theory that would avoid the philosophical embarrassments of Newton's theory.

That possibility was first envisaged clearly in the nineteenth century by Mach (1883). Mach's penetrating epistemological critique of the Newto-nian conceptions, in particular of the alleged connection between centrifu-gal force and absolute rotation, went beyond criticizing the epistemologi-cal shortcomings of Newton's theory; it showed the way to an alternative physics in which centrifugal forces, and inertial effects generally, would depend on the relation of a body to the other masses in the Universe. Like velocity in Newton's mechanics, rotation and acceleration in this new theory would be purely relative.

The task of fashioning these insights into a physical theory was left for Einstein, who absorbed Mach's ideas, but who had also absorbed the nineteenth-century revolution in the foundations of geometry; above all, Einstein understood the role of convention in connecting spatio-temporal concepts with empirical observation and measurement. He had already brought these ideas together in special relativity, giving an empirical defi-nition to simultaneity and, by the same stroke, revealing the meaningless-ness of Newtonian absolute simultaneity and absolute time: according to Einstein's criterion of simultaneity, simultaneity and time intervals must depend on the frame of reference. But special relativity had only shown the equivalence of uniformly moving frames of reference (inertial frames), and still maintained the distinction between these and accelerating or rotating frames. Mach's arguments showed Einstein that this distinction has no more epistemological legitimacy than the distinction between uniform motion and rest. Mach's idea then found a physical realization in the equivalence principle: because of the empirical indistinguishability of inertial motion and free fall in a gravitational field, the distinguished status of inertial motion and inertial frames could no longer be maintained. Thus the possibility of generalizing the special principle of relativity, from uniform motion to all states of motion, was realized. Physics had freed itself from the vestigial traces of Newton's metaphysics and had finally caught up with philosophy. A movement that was philosophically inevitable – the "relativization" of quantities naively thought to be absolute – was finally completed.

2.2 THE REVISIONIST VIEW

The foregoing is, more or less, the logical positivists' account of the history of the philosophy of space and time.[1] It faulted Newton not only for the details of his theory of space and time, but also for misunderstandings about scientific method, especially about the relation between theoretical entities and empirical facts. So it was, perhaps, unlikely to survive the later twentieth century's dissatisfaction with the positivists' own views on scientific theory and evidence. Nor could such an account survive the emergence, and gradual dissemination among philosophers, of a correction of the positivists' interpretation of relativity. Almost from the advent of general relativity, mathematicians and physicists who understood it particularly well, especially Weyl (1918, 1927) and Eddington (1918, 1920, 1923), expressed a very different view of the theory from Einstein's: not as a theory of the relativity of motion and the equivalence of frames of reference, but as a theory of the geometrical structure of the world (see Chapter 4, later). The essential feature of general relativity, on this view, was not that it eliminates the idea of a privileged coordinate system, but that it represents the geometry of space-time as a function of the mass–energy distribution. Therefore space-time is locally similar to the space-time of special relativity, but globally mutable and inhomogeneous. As a recent philosopher noted, commenting on some errors of the logical positivists, general relativity turned out to be "no less absolutistic about space-time than Newton's theory was about space" (Coffa, 1991, p. 196).

At the very least, we can identify a common metaphysical principle uniting general relativity with special relativity and Newton's theory:

space-time is an objective geometrical structure that expresses itself in the phenomena of motion. The theories disagree on *which* phenomena express that structure and precisely *how*; in general relativity the structure has the radically novel feature of being, not a fixed background, but a dynamical structure whose states depend on the states of the matter and energy within it. Eddington and Weyl, perhaps especially the latter, were quite emphatic about the tremendous philosophical significance of general relativity and of its departures from its predecessors. But what they emphasized (as we will see later) was clearly something completely different from the sweeping methodological and epistemological differences claimed by the positivists.

The geometrical way of thinking about space-time theory, as developed by Eddington and Weyl, was developed and maintained among specialists in relativity theory, and by the 1960s had attained a more or less standard form.[2] General relativity, special relativity, and Newtonian space-time could all be represented in a common mathematical framework, in which space-time is thought of as a differentiable manifold with geometrical structures defined by tensor fields; the various theories amount to differing choices of the tensor fields. But within the philosophy of science this view, and the weaknesses of the positivists' view, first came into prominence with the publication of Stein's "Newtonian space-time" (1967). Stein's interpretation of Newton brought out the historical and philosophical carelessness of the standard empiricist polemics against absolute space and time. In doing so, moreover, it revealed the deep connections between Newton's ideas about space and time and his dynamical theory. For application of the laws of dynamics – as understood not only by Newton, but by his foremost philosophical critics as well – was based on the analysis of particle trajectories, and so required a spatio-temporal framework that would suffice for the analysis of trajectories. The crux of Newton's argument for absolute space, then, was that this requirement could never be fulfilled by the Cartesian and relativistic views of space and time favored by his contemporaries. Therefore, in the literature since Stein's paper, Newton's theory is no longer regarded as a naive metaphysical appendix to his physics. Instead, it is regarded as a fundamental challenge to relationalism, one that the relationalists of Newton's time were very hard pressed to answer, and with which even relationalists of the present day must reckon. In this manner the absolute–relational controversy, which the positivists thought had been settled by Einstein, came to life once again.

This rehabilitation of Newton's philosophy was undoubtedly a change for the better, and it brought the philosophical debates concerning space-time into closer contact with the foundations of physics; it no longer seemed acceptable to argue against Newton on general epistemological principles, without regard for the presuppositions about space and time that may be required by physics. Yet, as we will see, the essential point of Stein's paper – and therefore of Newton's arguments – has not been fully appreciated. The contemporary literature assumes that Newton was trying to answer a standing metaphysical question – are space, time, and motion absolute or relative? – and that he brought physics to bear on this question much more convincingly than earlier philosophers, especially the logical positivists, had allowed. What this assumption overlooks is that Newton did not try to answer that question at all; on the contrary, he did not even take for granted that such a question was well-posed. For this reason Newton did not even attempt to show that space, time, and motion are absolute. His primary aim, instead, was to *define* "absolute space," "absolute time," and "absolute motion": to exhibit empirical criteria for applying the concepts, and to reveal the roles that they play in solving the problems of mechanics. The crucial secondary aim was to show that the corresponding concepts defined by his contemporaries, as purely relative notions, were for any mechanical purpose quite useless.

This interpretation of Newton is still considered eccentric, despite the prominence of Stein's paper for nearly four decades; indeed, in the large body of literature that cites his paper, this crucial aspect of it is rarely noticed (see DiSalle, 2002a for a more detailed account). But it is amply supported by the text of Newton's "Scholium" on space, time, and motion, and even more by his unpublished *De Gravitatione et aequipondio fluidorum*. In both texts, Newton's problem is never to justify metaphysical claims about space, time, and motion, but to define the concepts in a way that connects them with the laws of physics, and with the empirical practice of measurement. Accordingly, Newton's central argument against his contemporaries is directed against their definitions. What they attempt to define as the "philosophical" conception of motion is incoherent with the natural philosophy that they practice themselves.

2.3 THE SCIENTIFIC AND PHILOSOPHICAL CONTEXT OF NEWTON'S THEORY

Newton introduced his theory of space and time not in the body of the *Principia*, but in a Scholium to the preliminary "Definitions." This circumstance might already warn the reader that Newton is not about to answer already defined questions about space and time; instead, he is about to set aside the terms of the prevailing philosophical discussion of space and time, and to introduce theoretical concepts of his own.

Although time, space, place, and motion are very familiar to everyone, it must be noted that these quantities are popularly conceived solely with reference to the objects of sense perception. And this is the source of certain preconceptions; to eliminate them it is useful to distinguish these quantities into absolute and relative, true and apparent, mathematical and common. (Newton, 1726 [1999], p. 408)

As Stein was the first to emphasize (1967), the "preconceptions" Newton refers to are those of Descartes and his followers. The Cartesian approach to physics had (at least) two notable aspects whose connections with one another Newton found extremely problematic. On the one hand, as a program for mechanical explanation, Cartesian physics was an extension of Galileo's: its basic problem was to explain motion mechanically, and its fundamental assumption was that a body persists in a simple, uniform motion until external influences interfere. In Galileo's case, the assumption was that "nearly" uniform motions parallel to the Earth's surface – that is, circular motions – would persist, so that the rotation of the Earth must be nearly undetectable by mechanical experiments performed on the Earth; thus objects seem to fall in a straight line to the center of the Earth, in spite of the constant rotation, because their continuing horizontal motion is simply composed with their gravitation toward the Earth. Descartes and his followers extended this idea to account for all motion in the Universe, according to the principle that only *rectilinear* motion persists, and that every deviation from rectilinear motion – anywhere in the infinite Euclidean space of the Universe – requires some mechanical explanation. And a mechanical explanation, for them, necessarily involved the direct communication of motion from one body to another, by impact; no other sort of influence of one body on another could possibly be physically intelligible. To mention two of the most familiar examples, light was supposed to be a pressure propagated instantaneously through a universal medium, and the motions of the visible planets were supposed to be caused by the pressure of an unseen fluid that carried them along.

On the other hand, Cartesian physics came with a distinctive philosophical account of space and motion, an account that had a role to play in the program for mechanical explanation as well as in the philosophical interpretation of it. For Descartes, space and matter were essentially the same: material substance has no essential property but extension, and extension is obviously the essential property of space as well. Therefore, where there is extension there is also substance, by definition, and it is only our way of conceiving them that creates a distinction between the two. Hence Descartes' argument for the impossibility of a vacuum: empty space is impossible *by definition*, since wherever there is extension there is, by definition, substance. From this principle Descartes derived his mechanical explanation for planetary motion. Since the identity of space and body means that the Universe is necessarily full of matter, the only possible motions are circulations of matter about various centers; therefore the Universe consists of fluid vortices that carry systems of planets around their central stars, and systems of satellites around their central planets. Moreover, all such motion must be interpreted, from Descartes' philosophical point of view, not as motion with respect to space, but as motion with respect to the fluid medium. While the vulgar think of motion as "the action by which a body passes from one place to another," motion "in the philosophical sense" must be understood as the body's "transference from the vicinity of those bodies contiguous to it to the vicinity of others" (Descartes, 1983, p. 52). This definition appears to be motivated by the desire to assign an unequivocal state of motion to everything: there are "innumerable" motions in every body, depending on what other things we choose as a standard of reference, but Descartes' criterion assigns to each body one motion that is "proper" to it. Among all bodies relative to which a given body may be said to be moving, those that are immediately contiguous to it have a position that is, at least, unquestionably unique. Therefore the application of this criterion ought to be free of any ambiguity.

Newton saw that these two collections of principles – the program for mechanical explanation, and the philosophical account of space and motion – cannot stand together. The vortex theory explains planetary motion by assuming that the planets would move in straight lines, but for the fluid that carries them along, balancing the centrifugal tendency of each planet against the pressure of the surrounding medium. Therefore Descartes' *causal* account of the motions supports a Copernican or Keplerian model of the Solar System, sustained by the rotation of the Sun as it is communicated to the celestial matter. But his philosophical account of motion allows him to equivocate on the great question of the system of the world: since the Earth is being carried by the fluid, and does not move relative to the particles immediately surrounding it, the Earth is "philosophically" at rest. Thus, Descartes asserts, "I deny the movement of the earth more carefully than Copernicus, and more truthfully than Tycho" (1983, 3:19). This separation of the philosophical from the causal understanding of motion is what Newton found most problematic in Descartes' theory, and convinced him that natural philosophy could not proceed without proper definitions of space, time, and motion. The question addressed by the Scholium, therefore, is not whether space, time, and motion are "absolute." It is, rather, how the concepts of space, time, and motion must be *defined* in order to provide a coherent basis for dynamics.

2.4 THE DEFINITION OF ABSOLUTE TIME

Newton begins by defining absolute time: "Absolute time, without reference to anything external, flows uniformly" (1726 [1999], p. 408). Since this is not a metaphysical claim, but a definition, it makes no sense to ask the question that is traditionally asked, that is, whether Newton succeeds in proving it. The appropriate question is, instead, is this a good definition? Does it actually define any physically meaningful quantity? In fact, two concepts are involved in Newton's definition: absolute equality of time intervals ("uniform flow"), and, less obviously but equally essentially, absolute simultaneity. (See Figure 1.) Both are in fact necessary to the physics of the *Principia* – and, indeed, to all of seventeenth-century mechanics. Absolute simultaneity is the more pervasive concept, underlying as it does not only physics, but the notions of past, present, and future as understood at that time (and after, at least until special relativity). Even Leibniz, who claimed to reject absolute time, never doubted – on the contrary, central parts of his metaphysics required – the reality of the distinction between contemporaneous and successive events. In fact this distinction is implicit in the idea of the spatial order of things at a given instant, and in the idea of relative motion as the change of spatial distances between bodies from moment to moment. Within Cartesian physics, absolute simultaneity was implicit in the theory that light is the effect of pressure that is instantaneously propagated through the celestial medium, which implied that distant events may be perceived simultaneously with their occurrence. Indeed, the pervasiveness of the concept can be judged from the revolutionary character of special relativity: centuries of polemics on the "relativity of time" scarcely prepared anyone for the relativity of simultaneity.

The more problematic concept, at first glance, appears to be that of uniform flow. It is far from obvious what objective grounds could exist for judging that two intervals of time are truly equal. Any measurement of time intervals is necessarily based on the observation of some motion, presumably a periodic process; on what ground could we assert that a certain natural cycle truly repeats itself in equal intervals of time, or that any clock that we might devise can achieve or even approximate that ideal? Our ground is, simply, the laws of motion; the distinction between equal

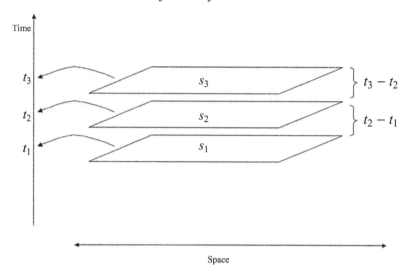

Figure 1. Newtonian absolute time: the world of space-time is the successive situations of space (s_1, s_2, s_3, etc.) at successive moments of time (t_1, t_2, t_3, etc.), and there is an objective measure of the ratios of time intervals ($t_3 - t_2$, $t_2 - t_1$, etc.).

and unequal time intervals is implicit in the distinction between inertial motion and motion under the influence of a force. In principle, equable flow is defined by the first law of motion: equal intervals of time are those in which a body not subject to forces moves equal distances. This definition was not given explicitly in these terms until Euler (1748), but it is evident in Newton's association of "truly equable motion" with motion that is "not accelerated or retarded" by any external force or impediment. Physics therefore provides us with a definition of absolutely equable flow, just to the extent that it provides us with objective criteria for measuring forces; the extent to which we can approach Newton's ideal is just the extent to which we can account for all the forces acting on a given body. It follows that the measurement of absolute time implicitly requires all three laws of motion. For only with the second and third laws do we have the criteria to distinguish genuine forces from merely apparent ones, and thereby to determine how closely any given motion approaches the ideal.

The empirical meaning of absolute time, in short, is that it licenses a line of approximative reasoning: it makes sense of the notion of *improving* the measurement of time to any arbitrary degree. In the case of simultaneity, this amounts to supposing that errors in the synchronization of spatially separated clocks could be, in principle, made arbitrarily small, or that signals informing us of distant events could be arbitrarily fast. In the case of equal time intervals, it amounts to supposing that a clock may be improved arbitrarily, yielding an increasingly good approximation to truly "uniform flow." In principle, such improvements involve more or less straightforward applications of the laws of motion, in the precise determination of all the forces that are at work in any particular physical process. But it was clear to Newton that any actual motion will likely fall short of the ideal. When we "correct" the time intervals measured by, say, the Earth's rotation, we assume that astronomical motions provide a better approximation to uniformity, being less subject to external disturbances. So, in effect, we judge the motion of the Earth by how well it corresponds to astronomical motions, that is, by how closely the intervals it measures approximate intervals of astronomical motions. In sum, if the laws of motion are true, they allow us to judge how well any actual motion realizes the ideal of inertial motion, and so to judge how well any cyclic motion – any clock – comes to measuring true time.

It should be clear, then, that Newton's theory of absolute time is entirely derived from fundamental assumptions shared by the mechanical philosophy: that there is a genuine physical distinction between inertial and non-inertial motion, and that there is an unambiguous way of determining all of the forces involved in every non-inertial case. So the objections raised against the idea at the time, coming from philosophers who shared these assumptions uncritically, stood on very questionable ground. In particular, the classic objections to absolute time raised in the absolute–relational debate – the Leibnizian indiscernibility arguments – are completely beside the point of Newton's discussion. The Leibnizian critique is based on preconceptions of the terms that Newton is using: if Newton is claiming that time is absolute, he must be implying that time is a substance, and for Leibniz real substances are, or are composed of, distinct individuals. No difference could be discernible, however, between our Universe and one in which all events were arbitrarily shifted forward or backward in time; for time is only an "order of succession," not a collection of moments that possess distinct individual natures. Such a shift would therefore be an empty distinction between things that are truly indiscernible (Leibniz, 1716, pp. 404–5). But Newton's definition does not imply any such distinction: the only distinctions that Newton's concept requires are between simultaneous and non-simultaneous events, and between equal and unequal time intervals. As Earman (1989) put it, absolute time is a theory, not of the *ontology* of time but of its *structure* (p. 8).

A more telling objection was the one raised in the nineteenth century, notably by Neumann (1870) and Mach (1883): if the Newtonian definition of equal times is in fact a definition, it is difficult to see how it can be anything more than a convention. To use a nineteenth-century example (see Thomson, 1884, p. 386), consider a fly buzzing about at random; can we determine objectively that its motion is not uniform? To do so, we must already have some standard of uniformity in hand. The laws of motion seem to provide one, insofar as particles that are free of all forces may be said to move uniformly. But how do we identify the free particles, if not by their uniform motion? Relative to the reference frame whose origin coincides with the center of gravity of the fly, the fly's motion must certainly appear uniform. From considerations of this sort, Neumann and Mach concluded that the first law of motion, as stated by Newton, is not really an empirical statement, since any motion may be conventionally designated as the uniform standard. According to Mach, we must assign some empirical content to the law by choosing the most obvious and convenient standard: we stipulate that equal time intervals are those in which the Earth turns through equal angles, and that all free particles travel equal distances in intervals in which the Earth turns through equal angles. According to Neumann, however, the first does have an empirical content once we apply it to more than one particle: that one particle moves uniformly is a stipulation, but it is an empirical claim that some second free particle moves, with respect to the first, equal distances in equal times. Then we can define equal times as those in which any two free particles move proportional intervals of time. Rather than a mere convention, Neumann's view states as a law of nature that all free particles will travel in straight lines, and the distances that they travel will be mutually proportional. (See Figure 2.) Thus his version expresses what absolute time really means in classical mechanics.

One might suspect, however, that the difference between these two versions is largely an illusion, and that as far as experience is concerned, they amount to the same thing. Even if the principle of uniform motion is stated in Neumann's form, any practical application of it will involve trying to find in nature – where, as Newton himself acknowledged, there aren't any free particles – some motions against which we can judge the uniformity of others. Singling out some particular motion might seem to introduce an element of convention after all. We have to choose the standard, it would seem, on no other grounds than simplicity and convenience. In the twentieth century, the logical positivists regarded this conclusion as a central feature of the revolution introduced by general relativity, and as one of the ways in which Mach was vindicated by Einstein. If we found it sufficiently simple – to use their famous example – we might designate the heartbeat of the Dalai Lama as the standard for equal time intervals. If we reject this

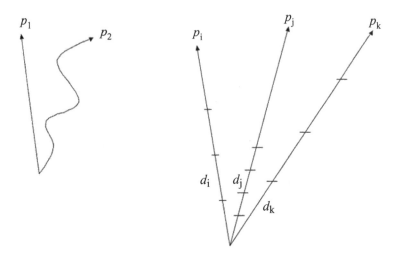

Figure 2. The definition of equal time intervals: either of p_1 or p_2 may be designated as a uniform motion by arbitrary stipulation. But it is an empirical claim that, for several particles p_i, p_j, p_k, the distances traveled are *mutually proportional*, i.e., that in intervals in which p_i travels equal distances d_i, p_j and p_k travel equal distances d_j and d_k.

choice, it is not because we have any way of determining that the intervals between beats are objectively unequal; the only reasonable complaint is that they make for an *inconvenient* definition of equal times. This inconvenience would reveal itself in the fact that few if any other motions would be proportional to them in Neumann's sense, even to some rough approximation. If the Dalai Lama's heartbeat were truly an ideal clock, it would have to be admitted to be the only one; it would be difficult if not impossible to construct another clock to agree with it, maintaining some approximately fixed proportion between the heart's beating and its own. As a result, it would be difficult to incorporate this measure of time into any simple or convenient system of laws of motion.

This account undoubtedly has an element of truth. Even if something like absolute time really exists, any empirical measurement of time requires the choice of some convenient standard. And if Newton's account really is, as I've suggested, only a definition, then a certain arbitrariness would seem to attach to it in any case. But the conventionalist view doesn't fully comprehend the connection between the definition of absolute time and the empirical content of Newton's laws. For, if the Universe is governed by those laws, then the correspondence between natural clocks is more than a matter of convention. In the ideal case, the laws assert that two inertially moving bodies, or ideally constructed clocks, will measure proportional intervals; in practice, they assert that the more nearly clocks approximate the character of ideal clocks, the more nearly they will agree with one another, and the more nearly the intervals that they measure out will be mutually proportional. Of course, in the case of a failure to agree, the conclusion that a clock is not running uniformly could always be evaded by some hypothesis or other about disturbing factors. From this we could conclude that the definition of equal times is still a definition, and its application must involve certain considerations of simplicity after all. The empirical claim contained in the theory of absolute time, then, is only that the line of approximation that it licenses will actually work: that time measurement will in fact improve, and our clocks will be in increasingly good agreement, to the extent that we construct them by correctly applying the laws of motion. Such clocks, moreover, will agree increasingly well with the least-perturbed motions found in nature. In a universe that is more or less obedient to Newton's laws, the claim (for example) that the rotation of the Earth is slowing down makes perfect empirical sense.

2.5 ABSOLUTE SPACE AND MOTION

We have seen that, despite the philosophical criticism that greeted it and has followed it for centuries, Newton's definition of absolute time is no more than an accurate analysis of what was presupposed concerning time, in the science of mechanics as Newton and his contemporaries practiced it. Even more controversy has attended Newton's conceptions of absolute space and absolute motion, and, to a great extent, the controversy is similarly misguided. For, here again, Newton attempts to defend definitions of space and motion rather than hypotheses about them. In this case, however, much of the controversy has to do with defects in the definitions themselves. To understand why, we have to approach these definitions as we approached the definition of time, and ask what genuine physical quantities Newton's analysis successfully defines.

According to Newton's definition, absolute space is "without regard to anything external, homogeneous and immovable"; its parts are the "absolute places," which "all keep given positions in relation to one another from infinity to infinity," and "absolute motion" is translation from one absolute place to another (Newton, 1726 [1999], pp. 408–9). In short, absolute space is that with respect to which the velocity of every body is its true velocity. It requires, therefore, that we should be able to say of any thing whether it occupies the same place from moment to moment. In

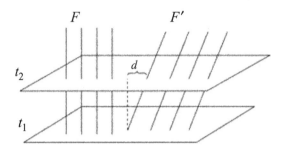

Figure 3. Absolute motion in absolute space: absolute space implies an objective distinction between motion and rest, i.e., between a family of parallel trajectories *F*, which pass through the same spatial positions at successive moments of time, and another family *F'*, which maintain the same mutual distances, but which change spatial positions over time. Any shift is allowed that would leave the trajectories *F* at rest and would maintain the spatial distance *d* traveled by any member of *F'* in a given interval of time.

other words, it implies that there is a set of trajectories in space-time that may be distinguished as the histories of particles that remain at rest. (See Figure 3.)

Just as in the case of absolute time, then, the familiar "indiscernibility" arguments against absolute space are not relevant to the concept that Newton is trying to define. For, if the theory of absolute time can be regarded as "structural" rather than "ontological," the theory of absolute space can be regarded in precisely the same way. The reason why this fact has been difficult to see, and why absolute space has therefore been the subject of so many misdirected arguments, has to do with another under-emphasized point of Stein (1967): Newton's theory of absolute space concerns the structure, not of space only, but of space-time. It implies not that "space is absolute" – whatever that might mean in Leibnizian or any other metaphysical terminology – but that *space is connected with time* in such a way that states of motion are well-defined. Arguments from the structure of space, then, like Leibniz's indiscernibility arguments, have no force against this theory. When Leibniz claimed that "space is something absolutely uniform" (1716, p. 364) – so that our universe is not distinguishable from one in which the position of every object is shifted in some arbitrary direction, or reflected from left to right – he was not really posing a relevant objection against Newton's theory of absolute space. Instead, he was merely pointing out some of the characteristic symmetries of Euclidean space that are fully incorporated into absolute space as Newton defines it. In particular, any shift in spatial position or direction is permitted as long as it leaves the velocities of bodies unchanged. In other words, Newton's theory only requires that the difference between motion and rest be respected, but does not in any way distinguish positions or directions in space. It must be possible to say, of any body, whether it occupies *the same* position over time, but *which* position it occupies is irrelevant. The Leibnizian arguments, in effect, confuse two issues: whether space allows for a distinguished position, and whether space-time allows for a distinguished velocity. The mere homogeneity of space, in fact, is completely independent of whether there are dynamically distinguished states of motion. For one could imagine an inhomogeneous space, in which it really would make a difference if one shifted everything to the left, say, into a region of increasing spatial curvature. Yet nothing could be inferred from that fact about the dynamical states of motion of bodies; that would require the further assumption of distinguished space-time trajectories. Of course neither Leibniz, nor anyone else in his time, could have contemplated the possibility of such a structure for space. But it does reveal something about his principle that "space is something completely uniform" that is generally overlooked in philosophical discussions of it: the principle is both a highly special assumption and, compared to what he was hoping to prove, extremely restricted.

Newton's definition of space, then, ought not to be judged against Leibniz's conception of substance. Instead it should be judged on its own terms, according to whether the distinctions that it implies have any empirical meaning. This, of course, is where the well-known difficulties arise. They arise, moreover, not from general philosophical principles, but from Newton's own dynamical theory. If dynamics provided some way to discern whether a body remains in the same place over time, or to measure its velocity, then absolute space would be empirically as well-defined as absolute time. In the seventeenth century, in spite of the currency of the Galilean–Cartesian principle of inertia, the idea persisted that the natural state of bodies is rest, and that force is required to maintain them in their motion; even Leibniz professed the latter principle, for he held that every body has a certain amount of "moving force" that, even though it is not empirically discernible, represents the body's true state of motion (see Leibniz, 1694, p. 184 and Chapter 3 later). But Newton had thoroughly embraced the principle of inertia as resistance to *change* of motion, and of acceleration as the true measure of force. Therefore he embraced the "Galilean relativity principle," that no mechanical experiment could possibly distinguish a system of bodies in uniform motion from one at rest. Evidently – even to Newton himself – this implied that no mechanical experiment could measure the velocity of a body in absolute space.

Indeed, few of his contemporaries – perhaps none other than Huygens – understood the Galilean relativity principle as clearly and explicitly as Newton. It is expressed at the very outset of the *Principia*, in Newton's definition of the "vis inertiae":

[A] body exerts this force only during a change of its state, caused by another force impressed upon it, and the exercise of this force is, depending on viewpoint, both resistance and impetus: resistance in so far as the body, in order to maintain its state, strives against the impressed force, and impetus in so far as the same body, yielding only with difficulty to the force of a resisting obstacle, endeavors to change the state of that obstacle. Resistance is commonly attributed to resting bodies and impetus to moving bodies; but motion and rest, in the popular sense of the term, are distinguished from each other only by point of view, and bodies commonly regarded as being at rest are not always truly at rest. (Newton, 1726 [1999], pp. 404–5)

In other words, Newton recognizes inertia as a Galilei-invariant quantity, so that impetus and resistance are the same thing seen from different points of view. On Leibniz's view, in contrast, the two differ fundamentally; what Newton here calls "impetus" is the "active" power of changing the state of motion of another body, while resistance is only a body's "passive" power to maintain its own state (Leibniz, 1699, p. 170). So uniform motion is a state of "activity," while rest is truly "inertial." In fact this distinction persists even in the thought of Kant.[3] Newton's discussion of inertia stands out as one that truly incorporates the principle of Galilean relativity.

Even more striking is that Newton derives the relativity principle explicitly as a corollary to the laws of motion: "When bodies are enclosed in a given space, their motions in relation to one another are the same whether the space is at rest or whether it is moving uniformly straight forward without circular motion" (Newton, 1726 [1999], p. 423). This is because those motions are determined by the forces of interaction among the bodies, and, by the second law of motion, these forces are entirely independent of the velocities of the bodies involved. Mechanics does indeed require, as Newton claims, a reference frame of places that "all keep given positions in relation to one another from infinity to infinity." But the laws of motion enable us to determine an infinity of such spaces, all in uniform rectilinear motion relative to each other, and the laws furnish no way of singling out any one as "immovable space." (See Figure 4.)

In modern terms, it is easy to explain what Newton should have said about this situation. A reference frame is the combination of two objects: a rigid spatial framework (places that "all keep given positions in relation

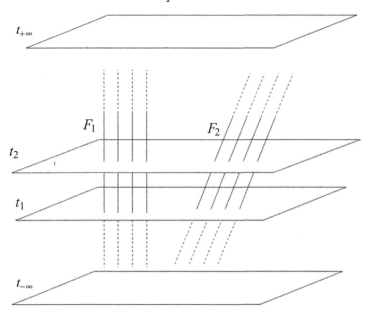

Figure 4. Galilean relativity: for any family F_1 of places that "all keep given positions in relation to one another from infinity to infinity," there is an infinity of other such families F_2, etc., in uniform rectilinear motion relative to F_1 and dynamically completely indistinguishable from it.

to one another from infinity to infinity") that allows us to describe motion purely kinematically, without regard to dynamical causes and effects; and some standard of time so that the motions can be expressed as a function of time. What Newton's mechanics requires, then, is an *inertial* frame: a reference frame in which bodies not subject to forces move in straight lines with respect to space and uniformly with respect to time; any acceleration of a body is proportional to, and in the direction of, an applied force; and every action corresponds to an equal and opposite reaction. Given one such frame, evidently, an infinity of others can be determined, each in uniform rectilinear motion relative to the first. It follows that any two (or more) inertial frames must be at rest or in uniform motion relative to each other.[4] Thus the forces, masses, and accelerations measured in one such frame will be the same in any other inertial frame, and any body that is rotating relative to one inertial frame will be rotating with the same velocity relative to any other. In other words, the shift from one such frame to another is just a change of perspective that does not alter any physically meaningful quantity. We can see this in the relations between two inertial frames with spatial and temporal coordinates (x, y, z, t) and (x', y', z', t'), the "Galilean transformations":

$$x' = x - vt$$
$$y' = y$$
$$z' = z$$
$$t' = t$$

where the x-axis is defined as the direction of relative motion. That the time coordinate remains unchanged reflects the fact that frames that agree on force, mass, and acceleration must agree on absolute time. But this conception of inertial frame, and therefore the solution to the problem of absolute space, did not emerge until the late nineteenth century (see Section 4.1 later).[5] The spaces referred to in Corollary V are in some sense equivalent to inertial frames, but conceptually quite different. For Newton's corollary does not specify these spaces intrinsically, by the properties of interactions within them, but extrinsically, by their states of motion (uniform and rectilinear) with respect to absolute space. To understand the inertial frame is to understand that its dynamical specification stands on its own, so to speak; it does not need to be supplemented by a larger space in which the frame is supposed to move. Newton was trying to articulate the dynamical conception of motion within a picture of space that was bound to make it more obscure.

The foregoing explains why, when Newton attempts to distinguish absolute and relative motion by their "properties, causes, and effects," he can only succeed partially. What he calls the properties of absolute motion in absolute space can be defined verbally, but they don't necessarily correspond to dynamical properties that are objectively measurable, and that do not depend on the arbitrary choice of an inertial frame. It is not surprising, then, that the conceptual analysis by which Newton defines the properties begins not from physics, but from common experience and discourse; it shows that the Cartesian definition of motion violates what appears to be the only reasonable way of thinking and talking about motion. Newton emphasizes three properties: that bodies at rest are at rest relative to one another; that parts of a body partake of the motion of the whole; that whatever is contained in a given space shares the motion of that space (Newton, 1726 [1999], pp. 411–12). Each of these is violated by motion in the Cartesian sense. All the planets, for example, are at rest in their own parts of the vortex, but they are not at rest relative to one another, and a planet may churn the vortical fluid by its rotation while its inner parts are considered to be at rest. Such a definition hardly seems to agree with any sensible use of the terms "motion" and "rest." But from the Cartesian point of view, there is a ready response to such a challenge: philosophy is simply not bound to ordinary usage, if it can arrive at a principled understanding of motion that is better suited to its aims. Descartes does indeed purport to offer such a principled understanding, insofar as his definition is supposed to provide a universal definition of motion; as was already noted, every body at every moment will indeed have one unique motion that is "proper" to it. The ordinary conception, by contrast, offers no way of discerning which of a body's myriad motions is its true motion. If Descartes' definition violates common-sense judgments – permitting us to believe, for example, that the pulp of an orange is at rest when the skin is in motion – this is only one more example of the ways in which the emergence of modern science must set common-sense judgments aside.

A proper philosophical response to Descartes, then, can only come from the causes and effects of motion, for only these provide for a principled empirical distinction with some foundation in the physics of motion. But because the empirically measurable causes and effects of motion have to do with inertia and force, they must respect the Galilean invariance of inertia and force. Therefore the causes and effects of motion can distinguish non-uniform motion from uniform motion, but never uniform motion from rest. So the empirical definition of absolute motion that Newton actually provides is more restricted than he had intended: he defines absolute acceleration and absolute rotation, but not absolute motion and rest. For instance, absolute and relative motion are to be distinguished by "the forces impressed upon bodies to generate motion" (Newton, 1726 [1999], p. 412). A body's relative motion may be changed without the application of any force; moreover, the forces that do act on a body do not necessarily change its relative motion, if it happens that the bodies relative to which we are measuring the motion are being acted upon by the same force. In the Cartesian case, the Sun's vortex acts on the Earth without changing its "motion in the philosophical sense," since the same particles of the vortex remain contiguous to the Earth. Newton, in contrast, defines true motion as that which cannot change without the action of a force, and which must change when a force is applied. By Newton's laws, however, a force can only be determined and measured by the changes it effects in the velocity of a body, changes that he shows (in the corollaries to the laws of motion) to be independent of the velocity itself.

Similarly, what Newton defines as the effects of absolute motion – centrifugal forces, or "the forces of receding from the axis of circular motion" – distinguish acceleration and rotation from uniform rectilinear motion. "For in purely relative circular motion these forces are null, while in true and absolute circular motion, they are larger or smaller in proportion to the quantity of motion" (Newton, 1726 [1999], p. 412). But they fail to distinguish uniform motion from rest, since the centrifugal forces produced in a body that accelerates or rotates correspond, not to change of place in absolute space, but to *change* or *difference* of velocity. The force experienced by an accelerating body is, again, independent of its initial velocity. In the case of a uniform and unchanging rotation, the centrifugal forces depend on the velocity of the rotation, but are independent of any translation of the rotating body through absolute space, i.e. of the velocity of the center of gravity of the body. Perhaps Newton assumed that, since the velocity of rotation is measurable from centrifugal forces, velocity in general is well-defined. But, as Huygens clearly understood – apparently alone, in Newton's time – centrifugal force is a function of a difference of velocity, namely the difference between the velocities of points on a rotating body; to take the simplest case, diametrically opposite points on a rotating disk have opposite velocities. This was Huygens' explanation why rotation, even though it involves no change of relative position, is a species of relative motion. (See Stein, 1967, 1977.) Unfortunately Huygens' account of rotation lay in an unpublished manuscript, and so Newton's understanding of the connection between space and absolute rotation prevailed, at least among those who accepted Newtonian physics, for two centuries.

In light of these difficulties with Newton's arguments, it is worth asking again that familiar question, what did he really accomplish in his famous "water-bucket experiment"? The experiment itself seems simple enough: suspend a bucket of water by a rope, and turn the bucket in one direction until it is "strongly twisted"; then, turn the bucket in the contrary direction and let the rope untwist. At first the bucket spins rapidly, but gradually the friction of the spinning bucket communicates its motion to the water. As that occurs, the centrifugal force increases until the surface of the water becomes concave and its outer edge climbs the side of the bucket. Stop the bucket, and the water continues to rotate, and its surface remains concave, until eventually the water also stops and returns to its initial state (Newton, 1726 [1999], pp. 412–13). Newton is thus drawing attention to the relation between the dynamical effects of rotation – the centrifugal forces – and how the rotation must be understood in Cartesian terms. When the bucket begins to rotate, the water begins to rotate in the Cartesian sense, i.e. relative to the bodies immediately contiguous to it. But its surface is, initially, flat. As the water begins to rotate, however, and to exhibit the centrifugal effect, it gradually achieves the same rotational velocity as the bucket. This means that as the dynamical effect is increasing, the motion in the Cartesian sense is slowing to a halt; when the dynamical effect is greatest, the Cartesian motion has ceased altogether. Then, when the bucket has stopped, the water continues to exhibit the dynamical effect, but now is suddenly rotating in the Cartesian sense once again. In other words, at the beginning and the end of the experiment, we see the same amount of Cartesian motion, but not the same dynamical effect, while in the middle of the experiment, the dynamical effect exists without any Cartesian motion. Newton has shown, then, that the Cartesian definition of motion has nothing to do with the salient dynamical feature of rotation, the "endeavor to recede from the axis of motion," from which "one can find out and measure the true and absolute circular motion of the water."

Therefore, that endeavor does not depend on the change of position of the water with respect to surrounding bodies, and thus true circular motion cannot be determined by such changes of position. The truly circular motion of each revolving body is unique, corresponding to a unique endeavor as its proper and sufficient effect . . . (Newton, 1726 [1999], p. 413)

In other words, even if the Cartesian definition identifies a single motion proper to each body, that motion is disconnected from the fundamental concerns of physics, especially the understanding of dynamical phenomena. As Newton shows, however, the dynamical phenomena themselves provide a unique measure of the state of rotation of any body. What is more, Newton points out that the concept of rotation that he is defining is the very one that the Cartesians employ, implicitly, in their attempt at a dynamical understanding of the Solar System. For in the vortex theory,

the individual parts of the heavens [i.e. of the fluid vortex], and the planets that are relatively at rest in the heavens to which they belong, are truly in motion. For they change their positions relative to one another (which is not the case with things that are truly at rest), and as they are carried around together with the heavens, they participate in the motions of the heavens and, being parts of revolving wholes, endeavor to recede from the axes of those wholes. (Newton, 1726 [1999], p. 413)

The crux of Newton's dynamical argument, then, is that the Cartesian definition ignores the aspects of motion that are central to Cartesian physics. It defines a univocal velocity for every body – indeed, every particle – in the Universe. But it does not offer any physical measure of the accelerations and rotations that are central to our understanding of the fundamental causal interactions.

But Newton's case for absolute rotation goes beyond criticizing Descartes' peculiar understanding of motion. In a second thought experiment,

Newton imagines two balls joined by a cord, revolving about their common center of gravity. Even if no other bodies are visible – even in the complete absence of any observable relative rotation – we could, he says, discern the rotation of the system. For "the endeavor of the balls to recede from the axis of motion could be known from the tension of the cord, and thus the quantity of circular motion could be computed." In other words, even if the system of two balls were entirely alone in the universe, its state of rotation could be known from the centrifugal forces. "In this way both the quantity and the direction of this circular motion could be found in any immense vacuum, where nothing external or sensible existed with which they could be compared" (Newton, 1726 [1999], p. 414). Where the bucket experiment was supposed to exhibit the futility of judging motion relative to contiguous bodies, this experiment is directed against the more general notion that rotation is relative. It defines a measure of rotation without reference to any relative standard at all.

This general claim is the one that has traditionally been the most controversial. Most of the controversy, however, has been based on a misunderstanding of Newton's purpose. One might, again, raise the objection that a sound dynamical definition of rotation cannot by itself provide a definition of absolute translation in absolute space. But the traditional objections are not directed to the superfluous character of absolute space, but to the perceived strangeness in the concept of absolute rotation. Objections of this sort were brought most forcefully in the arguments of Mach. Arguing along the same lines as in his criticism of absolute time, Mach held that every attribution of motion at least tacitly refers the motion to some physical object. If Newton finds the bucket an inadequate standard against which to judge the motion of the water, he is tacitly assuming that the fixed stars provide a better standard. If, furthermore, Newton is certain that the same centrifugal effects would exist even if the fixed stars were not present, he is extrapolating far beyond what can be justified by experience. All that experience can teach us about the laws of motion is how well they account for motion relative to the fixed stars. An empirically meaningful statement of the first law of motion, then, would be that a body not subject to forces moves uniformly, not "absolutely," but relative to "sufficiently many, sufficiently large and distant masses" (Mach, 1889, pp. 218–19).

Mach, at least, may be credited with understanding what Newton was really claiming. He recognized, that is, that Newton had defined the concept of rotation that is implicit in the laws of motion; his objection was to Newton's understanding the laws as something more than summaries of observed regularities. To speak of "absolute" rotation is to grant those laws an "absolute" status, asserting their general and abstract validity instead of seeing them for what they are: empirical generalizations about motion relative to the fixed stars, with the pragmatic and uncertain character that is typical of such generalizations.[6] The serious distortion of Newton's purpose came later, and is epitomized by Einstein's critique. In his argument for an "extension of the relativity principle," Einstein offers a now-familiar thought experiment: consider two spheres S_1 and S_2, rotating relative to one another, and suppose that S_2 bulges at its equator; how do we explain this difference? Einstein says,

An answer to this question can be acknowledged as epistemologically satisfying, only if the thing cited as a reason is an observable fact of experience . . . Newtonian mechanics gives no satisfactory answer to this question. It states the following: The laws of mechanics are fully valid in the space R_1, with respect to which the body S_1 is at rest, but not in the space R_2, with respect to which the body S_2 is at rest. But the privileged Galileian space R_1 . . . is a merely fictitious cause, and not an observable thing. (Einstein, 1916, pp. 8–9)

By this reasoning Einstein concluded that centrifugal and other inertial effects must be traced to some observable cause, such as the "distant masses," in order to remove the "epistemological defect" of absolute rotation. While this was an important motivation for general relativity, however, it is hardly an apt criticism of Newton. The objection assumes that Newton has introduced a privileged frame of reference – any of the spaces that are defined by Corollary V – as a causal explanation of the phenomenon of centrifugal force. But Newton is not invoking space as a cause at all. The cause of the centrifugal forces is the true rotation of S_1. But how does Newton know this? As we have already seen, Newton is simply presenting a *definition*: true rotations are by definition those that give rise to centrifugal forces. This is, again, the definition that is implicit in the laws of motion. Einstein's assertion about the laws of mechanics should therefore be rewritten: the laws of mechanics define an objective distinction between the space R_1 and the space R_2. Mach's doubt about the status of the laws of motion is not an unreasonable challenge to this definition, since the definition is obviously only as well-founded as the laws themselves. What Einstein called "Mach's principle," however – the positive *demand* that local inertial effects be explained by long-range interaction – is largely based on a misunderstanding. It may make sense as an alternative *hypothesis* to Newton's understanding of inertia, but not as a critique of its "epistemological defects."

2.6 NEWTON'S *DE GRAVITATIONE ET AEQUIPONDIO FLUIDORUM*

Einstein's misinterpretation of Newton expresses a common and continuing assumption of the absolute–relational debate. Even in discussions that are sympathetic to Newton's theory, it is treated as a hypothesis to be defended by dynamics, rather than as a set of definitions of the concepts implicit in dynamics. For this reason, Newton's manuscript *De Gravitatione*, since it was introduced to philosophers by Stein (1967), has been seen as an extremely significant work. In fact it is regarded as having provided something that the Scholium was lacking. For, on the usual view, the Scholium had only claimed that centrifugal forces can provide evidence of absolute rotation. *De Gravitatione*, however, contains a more general point: that Descartes' relativistic account of motion is completely incompatible with the laws of dynamics, at least as those were understood in the seventeenth century. For example, if the places and motions of the planets are defined in the Cartesian sense – that is, with respect to the immediately surrounding particles of the vortex – then it will be impossible to define a path for any planet, and therefore impossible to apply the most basic dynamical principles. Above all, as Newton emphasizes, it will be impossible to say of any body that its motion, insofar as it is free of applied forces, must be rectilinear and uniform. In other words, if it is impossible to define a privileged set of trajectories, the past and present state of a body will not suffice to determine its future states. In that case the entire explanatory program of classical dynamics will be impossible.

It follows that Cartesian motion is not motion, for it has no velocity, no definition, and there is no space or distance traversed by it. So it is necessary that the definition of places, and hence of local motion, be referred to some motionless thing such as extension alone or space insofar as it is seen to be truly distinct from bodies. (Hall and Hall, 1962, p. 131; see Stein, 1967)

Like the dynamical arguments of the Scholium, this argument infers more than the dynamics can justify; again, given a space that satisfies Newton's requirements for a dynamical description, any space in uniform motion relative to it will be equally satisfactory. Nonetheless, his argument makes it clear that dynamics requires a more elaborate spatio-temporal structure than the Cartesians and other relativists were willing to countenance.

Stein remarked that "if Huygens and Leibniz . . . had been confronted with the argument of this passage, a clarification would have been forced that could have promoted appreciably the philosophical discussion of space-time" (Stein, 1967, p. 186). In fact, no one made such an argument in public until the middle of the eighteenth century, when Euler presented a version of it in his "Reflexions sur l'espace et le temps" (Euler, 1748). Euler recognized that the Leibnizian theory of space crucially lacks the notion of *sameness of direction* over time:

For if space and place were nothing but the relation among co-existing bodies, what would be the same direction? . . . However bodies may move or change their mutual situation, that doesn't prevent us from maintaining a sufficiently clear idea of a fixed direction that bodies endeavour to follow in their motion, in spite of the changes that other bodies undergo. From which it is evident that identity of direction, which is an essential circumstance in the general principles of motion, is absolutely not to be explicated by the relation or the order of co-existing bodies. (Euler, 1748, p. 381)

As we will see in the next chapter, at least one Leibnizian was swayed toward Newton's view by this argument, namely Kant. But there can be no doubt of its impact on the philosophical discussion of space-time in the late twentieth century. When Stein brought this passage to the attention of philosophers in 1967, the character of the absolute–relational debate was changed. Stein's discussion was taken to have shown that, beyond providing some dynamical evidence for his own theory of absolute motion, Newton had presented a fundamental challenge to any relationalist view: to provide a dynamical theory that, without the use of any "absolute" structure – in particular, without an affine structure of privileged trajectories – can nonetheless explain and predict actual motions.

In spite of the prominence of *De Gravitatione*, however, some of its most important aspects have been largely overlooked. For one thing, it shows much more explicitly than the Scholium that Newton was not a "substantivalist," at least not in the now-standard use of the term. That is, he did not believe that space and time are substances, or that the points of space or the moments of time have distinct individual identities. Instead, he claims, "The parts of duration and space are only understood to be the same as they really are because of their mutual order and position; nor do they have any hint of individuality apart from that order and position which consequently cannot be altered" (Hall and Hall, 1962, p. 126). In fact, he expresses skepticism about the classical ontology of substance and accident – the kind of skepticism that would have been foreign to Descartes and Leibniz, and that only became current in philosophy with the work of Berkeley. In his attempt to understand the nature of material substance, Newton rejects the idea of "material substance" as the subject in which physical properties inhere. Instead, he proposes that a body is constituted by a collection of properties distributed over a region of space. Like Descartes, Newton reasons about what God would have to do in order to create such a world as we observe; the most natural possibility, Newton suggests, is that God "endowed" regions of space with the property of impenetrability. This theory of matter has been taken as confirming Newton's substantivalism about space; the principle that God can assign properties of matter to points of space appears to make the points of space "irreducible objects of first-order predication," and therefore individual substances in just the sense that Leibniz found objectionable. I don't believe that this interpretation can be sustained, however. The fact that the collections of properties that constitute bodies are assumed to be *moveable,* and to interact with one another according to the laws of motion, implies that their creation in no way endows particular parts of *space* with individuality. On the contrary, their unhindered mobility effectively *requires* that the parts of space are indistinguishable, and receive these collections of properties indifferently. Provided that their motions obey the laws of motion as we know them, these collections of properties will be indistinguishable from bodies as we know them. In this manner, Newton claims, we can replace the "unintelligible" notion of material substance with an intelligible one (Hall and Hall, 1962, pp. 139–40).

My point here is not to defend Newton's theory of material substance.[7] Nor is it merely to set the record straight about where Newton's views would stand in contemporary debates. Rather, it is to emphasize something distinctive about Newton's approach to space-time ontology. For Descartes or Leibniz, who had relatively settled views on the nature of substance and other ontological categories, it was important to determine where space and time stand in relation to these categories. Leibniz, in particular, underwent a serious intellectual change in his early career, from the quasi-Aristotelian view of space as a substance to his mature and more familiar view of space and time as mere "phenomena." It seemed clear to him, in any case, that if space and time could not satisfy his conception of substance, then their existence must be merely ideal. To Newton, in stark contrast, the traditional philosophical category of substance was not very well-defined. Indeed, he saw that it was difficult to make sense of the notion of material substance without appealing to space. Space itself, then, was neither substance nor accident, but had "its own manner of existing which fits neither substances nor accidents" – and therefore, to understand its peculiar manner of existing, he proposed to examine the roles that the concept of space plays in our actual knowledge of the physical world.

This brings us to the other largely overlooked aspect of *De Gravitatione.* Even more clearly than the Scholium, it shows that Newton was fundamentally concerned with the *definition* of space and motion, and that his fundamental argument against Descartes was a dialectical one, exposing the concepts of space and motion that are actually at work in Cartesian physics. In particular, Newton shows that there are in effect two "philosophical" definitions of motion at work in Descartes' physics: one is the aforementioned "motion in the philosophical sense," and the other is motion as it is understood in natural philosophy, that is, in the causal explanation of motion. Descartes calls the latter "motion in the vulgar sense," since it is referred to space or "generic extension" rather than to contiguous bodies. But it is clearly this latter sort of motion that makes a real difference to any physical and causal account of the planetary system. In Descartes' own account, the motion of the planets and comets in the vortex of the Sun is determined by the balance of their centrifugal tendencies against the pressure of the ambient fluid. Therefore natural philosophy ought to ignore the irrelevant conception of motion, and to adopt that conception that it requires to make physical sense of the phenomena. "And since the whirling of the comet around the Sun in his philosophical sense does not cause a tendency to recede from the center, which a gyration in the vulgar sense can do, surely motion in the vulgar sense should be acknowledged, rather than the philosophical" (Hall and Hall, 1962, p. 125). Given the general stridency of *De Gravitatione*'s attack on Descartes, the modesty of his language here is striking. He does not claim to have proven that motion is absolute, but only to have replaced a worthless definition with a sound one.

2.7 THE NEWTONIAN PROGRAM

If this interpretation of Newton is correct – as the texts of the Scholium and *De Gravitatione* seem to show beyond any doubt – it is important to reflect on why the traditional interpretation has persisted, particularly in the body of recent literature that rejects the positivists' account and acknowledges the influence of Stein. I see two chief reasons for this. First, we have inherited the absolute–relational debate, so to speak, from a tradition going all the way back to the Leibniz–Clark correspondence. Newton's Scholium, despite its original intent, has unquestionably played a major part in this controversy, and greater philosophers than ourselves have mistaken its definitions for outlandish metaphysical claims. Not merely the weight of tradition, but also a sense of the intrinsic interest of the questions involved – not only for the history of philosophy, but for the present and future of physics – understandably encourages the tendency to see Newton's arguments only as possible moves in this debate.

Second, and perhaps more important, is our understanding of Newton's metaphysical concerns, and their intimate connections with his views on theology as well as physics. If we grant that Newton was only defending definitions, it might seem that we are divorcing his theory of space and time from his profound beliefs about the real world, and making him out to be some kind of precursor of logical positivism – interested only in defining a useful conceptual framework, rather than in understanding the real arena in which God and human beings, as well as matter and forces, exist and act. That his arguments concern definitions, rather than making metaphysical claims, might seem incompatible with what seems to be an incontrovertible fact, namely, that Newton's views of space and time are essential to the ontological basis for his theory – part of his understanding of how the world really is.

This incompatibility is only apparent. It comes from the assumption that such definitions are arbitrary, as the positivists suggested, and adopted because of the simplicity and general usefulness of the conceptual framework in which they occur; it overlooks the fact that Newton's definitions emerge from a conceptual analysis. Newton undoubtedly believed in a world of real things, including God as well as material objects, things whose real causal interactions are governed by the laws of nature, whatever those might be. He therefore attempted to define space, time, and motion in such a way that this picture of the world might make sense – that is, not to stipulate the character of space and time, but to discover, by analysis of what we do know about the world, how they *must* be defined in order to make sense of such a world. In other words, in laying down these definitions Newton was not merely proposing a *possible* conceptual framework, but trying to identify the *necessary* framework, i.e. the concepts necessarily presupposed by mechanics as he and his contemporaries understood it. The fact that his attempt was not entirely successful – that, in the case of absolute space, he introduced a concept that was superfluous to mechanics – does not change the essential character of his argument: it is in fact a kind of transcendental argument, seeking to discover the conditions of the possibility of natural philosophy as practiced in his time.

This is an aspect of Newton's approach that his contemporary critics, the mechanical philosophers, never understood. The mechanists supposed that they had a similarly powerful argument for their program for physical explanation, in which all interaction was to be reduced to the exchange of momentum by direct impact. From their point of view, the reduction of interaction to impact was a condition of the possibility of understanding it at all. But this is an entirely different sort of argument from Newton's "transcendental" argument. As Newton and his followers (especially Cotes) emphasized, the mechanistic theory is no presupposition of scientific reasoning about motion; as the *Principia* itself shows to the contrary, the laws of motion could be successfully applied completely independently of any assumptions about the ultimate physical basis of interaction. As Newton emphasized, we have no philosophical insight into the physical basis of impact, but only some empirical rules that it appears to follow. Indeed, as far as the underlying nature of the interaction is concerned – what it is in the "essential" properties of bodies that makes the interaction possible – we know as little in the case of impact as in the case of gravitational attraction. And we know even less regarding the universality of the rules of impact: "the argument from phenomena will be even stronger for universal gravity than for the impenetrability of bodies, for which, of course, we have not a single experiment, and not even an observation, in the case of the heavenly bodies" (Newton, 1726 [1999], p. 796). This is an extremely significant remark, one whose significance was, arguably, not really appreciated by any of Newton's philosophical readers before Kant (see Chapter 3). For it concerns the entire post-Aristotelian understanding of motion, as something subject to *universal* laws rather than to two separate sets of laws, one for celestial and one for sublunary bodies; it reminds the reader that the argument for universal gravitation is the first real application of the laws of motion to the celestial realm. Rather than a departure from the program of modern physics, universal gravitation was, for the seventeenth century and a long time after, the only evidence that modern physics had the explanatory scope that the mechanists claimed for it.

If interaction by impact seems pre-eminently intelligible, then, it is not because we have penetrated into its inner essence. Rather, it is because we have brought its observable characteristics under the control of mathematical laws. Far from being a transcendental condition on physical inquiry, then, the mechanistic principle is inescapably hypothetical: it amounts to a hypothesis about the ultimate outcome of inquiry, but is in no way a precondition for the inquiry itself. On the basis of such a hypothesis, one might feel dissatisfied with a theory that fails to provide a mechanistic model, as Huygens, Leibniz, and their followers were dissatisfied with universal gravitation. But this dissatisfaction acquires its force more from a subjective hypothesis about the ultimate nature of reality, than from any inherent defect in the Newtonian program. Compared with the mechanical philosophy, that program is inherently more modest in its presuppositions – necessarily so, since its presuppositions are less restrictive than those of the mechanists concerning what form physical interaction must take. For in the Newtonian view, any interaction is physically intelligible as long as, and just to the extent that, it conforms to the laws of motion. The distinctive feature of Newton's program is precisely the careful separation of what physical inquiry must presuppose, in order to bring the actual motions within the grasp of the laws of nature, and what can be left open as an empirical question.

Newton did not share Leibniz's confidence in the power of metaphysics to grasp the reality underlying the physical world. Nor, of course, was he able to anticipate Kant's Copernican turn, and so to abandon the aim of comprehending things as they are in themselves. And he was hardly in a position to anticipate Carnap and the logical positivists, and to treat physics as merely a convenient conceptual framework founded on arbitrary definitions. But he did have a remarkably clear and self-conscious belief that, if there is to be any knowledge of space and time as they are in themselves, it must come from the ways in which the concepts of space and time function in our empirical knowledge. Newton shared with the mechanical philosophers the idea that, in the new science, the concepts of space and time, motion and causality, were explicated in novel ways. Where he parted from his contemporaries was in his belief that these novel explications came from the science itself – that their authority rested, not on their conformity to epistemological and metaphysical principles with which mechanists wanted to associate science, but on the roles that they play in the laws of science. What must be the nature of space and time, in order for the world to be as it appears to be, and to follow the natural laws that it appears to follow? This is Newton's question. For this reason, his view of space and time is not hypothetical in the way that Leibniz's is. Newton's theory is not, in other words, a hypothesis about the nature of space and time, like the hypothesis that they are mere phenomena deriving from an underlying world of monads and their perceptions. If Newton has a fundamental hypothesis, it is, rather, that the world is really governed by the laws of physics as then understood – by Leibniz and Huygens as well as by himself. This is why his arguments must be dialectical arguments, whose premises are only the common knowledge and assumptions of all serious physicists in his time.

Given all of these considerations, Newton's task now seems to have been a relatively straightforward one. That is, his readers evidently did not need to be persuaded of the fundamental principles of physics; they only needed to confront the incompatibility between those principles and their philosophical ideas about space, time, and motion. It is instructive to compare this situation to that of Galileo, attempting to persuade his readers to adopt such principles in the first place – that is, to abandon the Aristotelian conception of natural motion, and to accept something like the Newtonian principle of inertia. This situation appears, at least at first glance, to be a much better example of a Kuhnian conflict of paradigms: Galileo is attempting to persuade his readers, not merely to abandon some particular principle, but to think of a fundamental concept in an entirely new way. Again, this would seem to be the kind of argument that Kuhn would have said is "necessarily circular" (Kuhn, 1970a, p. 94), precisely because it advocates a definition rather than a proposition that can be assessed by normal scientific means. But Galileo's argument, we can see, has essentially the same dialectical structure as Newton's. Without satisfying the canons of normal science in Kuhn's sense, then, Galileo's argument nonetheless has an objective impact on the Aristotelian conception, for it throws an objective light on the relations between the professed doctrines of the Aristotelians and the assumptions that tacitly guide their empirical practice.

The classical arguments against the motion of the Earth appeared to have a sound basis in the theory of natural motion, which rested on the division of motions into natural and violent: heavy bodies naturally fall directly toward the center of the Earth, light bodies naturally rise from the center, and bodies that are neither heavy nor light – the celestial bodies, made of the "fifth element" rather than the earthly elements – revolve around the center. It would follow that, if a stone were dropped from a tower, it would naturally seek the center of the Earth. If the Earth were rotating, a stone dropped from a tower would fall behind the tower, continuing its vertical fall to the Earth as the tower continued horizontally. But, as Galileo pointed out, the vertical fall of the stone is not a fact, but an interpretation: if the Earth is at rest, the fall is vertical, but if the Earth is rotating, then the actual motion must be *composed* of its vertical motion and the horizontal motion that it shares with the Earth and the tower. So the argument from the vertical motion of the stone, by itself, is circular; its force depends on another assumption, namely that motions cannot be composed in the way that Galileo suggests, because it is impossible for the horizontal motion to persist without an external cause (Galileo, 1632 [1996], pp. 149–51). The Aristotelian theory explains the phenomena of motion, then, by categorizing them according to the various essences of bodies, and the essence of a heavy body is manifestly incompatible with the persistence of horizontal motion.

This fairly obvious point about Aristotle's theory helps us to identify exactly where Galileo's theory demanded a crucial change of perspective. From Galileo's perspective, a conception of natural motion is the foundation for an explanatory theory. As such it raises a characteristic type of question, namely about why any particular motion deviates from its natural motion in just the way, *and to just the degree*, that it does. The most familiar example is Galileo's explanation of the motion of a projectile as the composition of its gravitational free-fall and the inertial motion that persists after the moment of projection. At one level, then, such a theory of natural motion defines the basis for a quantitative theory of the violent motions and the forces that produce them; it provides the general form of any explanation in precise mathematical terms. And one might offer a plausible Kuhnian argument that such a fundamental change of perspective could never be justified on rational grounds. For the kind of explanation that Galileo's method promises is simply not an Aristotelian concern; the very appeal to this promise, as an advantage over the older view, marks a change in the very nature of science – a change in epistemic standards that inevitably results in a clash of incommensurable views.[8]

But Galileo is also arguing at a deeper conceptual level that is independent of the requirements of a quantitative theory of motion (which he himself, in any case, had not fully developed). He is arguing that the traditional concept of natural motion, as it is applied even in pre-theoretic practice, does not quite make sense – or, rather, that the Aristotelian way of making sense of it, as a philosophical concept, is incoherent with the practice of applying the concept in some very familiar circumstances.[9] In ordinary experiences of relative motion, Galileo points out, empirically sound judgments are those that ignore the Aristotelian conception, and implicitly assume that motion naturally persists when an applied force has ceased to work. When a rider on horseback wishes to throw a javelin in the air and catch it again, he does not try to allow for the velocity of the horse; instead of throwing ahead of himself, he must throw directly upwards just as if he were at rest (Galileo, 1632 [1996], p. 165). When a shooter wishes to hit a moving target, he does not attempt to "lead" the target, but instead "follows" the target with the barrel of the gun, implicitly knowing that the motion imparted by the moving barrel will be simply composed with the motion imparted by the powder charge (Galileo, 1632 [1996], p. 187). It never occurs to travelers on a smoothly moving ship to drop a stone from the mast, to test the theory of the persistence of motion, but no more does it occur to them to try to compensate for the velocity of the ship in their own ordinary movements (Galileo, 1632 [1996], pp. 195–7). From all of these examples, it appears that the concept of motion is not one that can be fully analyzed without considering its application to states of relative motion. When Galileo does consider its use in these contexts, he uncovers an implicit notion of the persistence of motion, and of the composition of horizontal and falling motions, that conflicts with what is explicitly professed.

As is widely known, Galileo was not advocating, nor evidently did he grasp very clearly, the precise association of force with change of momentum or the precise "Galilean principle of relativity" as we understand it. These arose from the combined work of Huygens, Newton, and others. This should not be surprising, given that the conceptual analysis that he undertakes does not begin, as it did for Newton, from a body of precise experimental knowledge in which the new conception of force – including the principle of action and reaction as well as the precise principle of inertia – was applied in quantitative detail. Rather, the starting point of Galileo's analysis was only a body of commonplace, little-controlled, essentially pre-systematic experiences of motion near and parallel to the surface of the Earth. It is therefore only appropriate that Galileo's conception of inertia implies the persistence of uniform circular motion; that is a reasonable extrapolation from the kind of experience that he is attempting to analyze. For the conclusions that we can draw from experience of this sort are not sensitive to differences that would appear on a larger scale, i.e. between uniform rectilinear motion and motion parallel to the Earth's surface. Yet from this starting point, a familiar experience such as the motion of a projectile becomes a well-defined problem of mathematical physics, namely, to derive the actual path of the projectile as a composition of its natural inertial motion with the accelerated motion produced by gravity.

Unlike Newton, then, Galileo had to defend a novel starting point, and to show that the new definition of natural motion was not as alien to common sense as Aristotelian philosophy had made it seem. The use of the dialogue form, then – at least as far as this definition is concerned – was no mere literary device or rhetorical move on Galileo's part. It reflects his profound understanding of the way in which his new definition emerged from ordinary experience of motion and force, and, consequently, of the fact that any objective justification for it must be dialectical in character. Moreover, it reflects a deep insight into the nature of dialectic itself. Plato thought that the mathematical sciences must fix a starting point arbitrarily, so that their principles must always be ultimately hypothetical, and their certainty relative. Dialectic was held to be a process unique to philosophy, in which, instead of setting down basic principles for the purpose of deductive argument, we might "ascend" to the truly fundamental principles by another sort of argument altogether (see *Republic*, Book VI). By its means, we were supposed to be able to "recollect" what is present in the mind but obscured by experience, namely the Forms, which constitute the genuine reality underlying the confusing appearances. Galileo also intended his dialectical argument to arrive at a fundamental principle, and explicitly compared it to Platonic recollection of something that we already know (Galileo, 1632 [1996], pp. 200–6). But for Galileo, what we "recollect" in the process is not some notion that transcends our ordinary experience, but the concept that implicitly guides our experience – even when our systematic, reflective interpretation of such a concept is something else altogether. In this respect it is perhaps a more self-conscious account of dialectic than that of Plato, who represented dialectic as an approach to the transcendent, but whose practical application of it – like Galileo's – usually appealed to common empirical knowledge and practice. It is no accident that both Plato and Galileo commonly appealed to the experience and practice of artisans and other practical people; this was in fact the pre-eminent source of knowledge that is systematic but not explicitly formulated nor, therefore, seen clearly in its relation to our more explicit philosophical or scientific beliefs. This is why it is just the sort of knowledge whose "recollection" can eventually lead us to revise our more explicit beliefs.

The dialectical way of arguing that Newton used, then, had an established use in the unfolding of the "scientific revolution." Apart from the familiar rhetoric in favor of drastic methodological change – the demands for more experiment and observation, for inductive methods, for more precise quantitative reasoning, and so on, that we usually associate with the advocates of early modern science – the transformation of science in the seventeenth century required a defense of certain drastic conceptual changes. To the later twentieth century, it seemed that any such defense must have been a circular argument: first, because it was difficult to see how it could have been the kind of inductive or hypothetico-deductive argument that the rhetoric of the scientific revolution claimed to rely on; and second, because it was hard to see how a more philosophical sort of argument could convey anything more than the subjective philosophical preferences of its author (see Kuhn, 1970b, p. 6). That is, it seemed impossible to represent such arguments as empirical and scientific, and therefore to represent them as making any significant contribution to the rational motivation for a drastic conceptual change. But once we recognize the dialectical dimension of conceptual arguments like those of Galileo and Newton, we are in a better position to appreciate their empirical dimension. They introduce novel concepts by a dialectical form of conceptual analysis, which demands some reflection on concepts that are used unreflectively in established empirical and scientific practices; they overcome traditional resistance by revealing that the novel concepts, in some latent form, are already in use in the best scientific reasoning of the time. As we can see from Newton's arguments about space and time, and about the limitations of the mechanical philosophy, this sort of conceptual analysis is fairly typical of his philosophical engagements with his contemporaries. He defends what appear to be his most radical notions, by revealing their basis in principles that his opponents already accept and use in their own reasoning.

2.8 "TO EXHIBIT THE SYSTEM OF THE WORLD"

The ultimate object of Newton's dialectical argument – "the aim for which I composed" the *Principia* – is to resolve the question of "the frame of the system of the world." And the thrust of the argument is that the accepted principles of mechanics contain, implicitly, a definition of true motion by which the question is radically transformed. To Leibniz, for example, the question was transformed by the philosophical insight that motion is purely phenomenal and relative: the question therefore has no objective answer, and we can do no more than choose the simplest hypothesis about which body is at rest. But the latter is not a completely novel idea; Copernicus himself may be said to have endorsed a similar view, when he offered only plausible arguments for placing the Sun in the center, invoking only the simplicity that this choice introduces into the structure of the system and astronomical calculations. Arguably, it was even present already in Ptolemy, whose arguments are essentially of the same form as Copernicus' – unsurprisingly, since Copernicus' defense of his hypothesis is closely modeled on Ptolemy's. But Newton's argument leaves no room for a hypothesis about the structure of the system; the only hypotheses are that the planetary system is a system of masses that interact according to the laws of motion, and that their relative motions are more or less as agreed upon by astronomers. Specifically, he is willing to assume that the planets, other than the Earth, orbit the Sun in conformity with Kepler's second and third laws; that the known satellites orbit their respective planets by the same laws. Then, by the laws of motion and their corollaries, he is able to deduce properties of the central forces, and eventually to compare the masses of those bodies that have satellites. In thus setting out the problem, Newton carefully leaves open the question whether the Earth orbits the Sun, or the Sun the Earth: his Phenomenon 4 states, "The periodic times of the five primary planets and of either the sun about the earth or the earth about the sun – the fixed stars being at rest – are as the 3/2 powers of their mean distances from the sun (Newton, 1726 [1999], p. 800).

But these fairly uncontroversial premises lead to an unequivocal result. For now the outstanding philosophical questions about motion have been redefined as empirical questions. The great question of the day – what is really at rest in the planetary system? – now has to be understood as a question about the center of gravity; since only the center of gravity is unaffected by the interactions of the bodies in the system, we can only ask which of the bodies remains closest to the center of gravity. Therefore, Ptolemy, Copernicus, Tycho, and Kepler are all wrong; neither the Earth nor the Sun is at rest in the center. And it is not merely that the question that they asked has been changed into an empirical one, rather than one to be answered by the most plausible hypothesis; the answer turns out to be an entirely contingent one, depending on the relative masses and their distances. Newton acknowledges that Kepler's view is closest to the truth: "if that body toward which other bodies gravitate most had to be placed in the center . . . that privilege would have to be conceded to the sun" (Newton, 1726 [1999], p. 817). Yet it is clear from Newton's reasoning that only the peculiar arrangement of the system – with most of its mass contained in the Sun – permits one of the traditional views to be even approximately correct. If the masses were more evenly balanced, the traditional question might make no sense at all.

If we return to the problem of absolute space, we can now see that the question, "is space absolute?" is not well-posed; the proper question is, does absolute space, as Newton had defined it at the outset, have some legitimate function in his program of explanation, like the functions of absolute time, absolute acceleration, and absolute rotation? The answer is clearly negative. But we can see from his account of the Solar System how thoroughly Newton grasped this fact. The theory of absolute space, simply put, has no role to play in his program. This is not merely a rational reconstruction; Newton himself explained that his program "to collect the true motions from their causes, effects, and apparent differences" has no use for the distinction between uniform motion and rest. For that program turns on identifying the center of mass for a system of interacting bodies and the forces of interaction among these bodies. By Corollary V to the laws of motion, that determination is completely independent of the velocity of the center of mass. Moreover, Corollary VI implies that it is independent even of the acceleration of the center, provided that the force causing the acceleration acts equally, and in parallel directions, on all parts of the system.

There are two kinds of application of this principle. One is the analysis of sub-systems within the Solar System, namely individual planets with their satellites, in which it is assumed that the action of the Sun on the parts of the system (say, on Jupiter and each of its moons) is so nearly equal and parallel that it may be neglected altogether, and the system treated as if it were isolated and moving inertially. Another is the study of the Solar System as a whole, which is independent of any acceleration of the system as a whole. Newton thus recognizes that his solution to the "frame of the system of the world" is extremely restricted: it can say nothing about whether the entire system is at rest, uniformly moving, or even uniformly accelerating under the influence of some yet-unknown external force.

> It may be alleged that the sun and planets are impelled by some other force equally and in the direction of parallel lines; but by such a force (by Cor. VI of the Laws of Motion) no change would happen in the situation of the planets to one another, nor any sensible effect follow; but our business is with the causes of sensible effects. Let us, therefore, neglect every such force as imaginary and precarious, and of no use in the phenomena of the heavens . . . (Newton, 1729 [1962], p. 2:558)

Newton's recognition of this fact reveals that his idea of determining the true motions, or "solving the frame of the system of the world," is free of any delusions about determining the velocities of bodies in absolute space.

Thus what Newton really means by "true motion," in his program to "determine the true motions," is precisely analogous to what he means by true time: it is an extrapolation from an empirically well-defined process of approximative reasoning. The ideal is a complete dynamical account of the accelerations in any system of bodies, in which every acceleration of every body in the system is part of an action–reaction pair involving some other body within the system; from these interactions the masses, and the position of the center of mass, will be known. The ideal case is therefore a system in which the true accelerations are known, and the absolute velocity of the center of mass is unknowable and irrelevant. (Again, by Corollary VI even the true accelerations need not be known, since the accelerations within the system may be composed with accelerations originating outside the system – provided that they affect all bodies in the system equally – without affecting the dynamical account of the system; this means that the "ideal" ideal case, in which we know the true accelerations, must be the case in which all bodies in the universe are comprehended in a single system, for only then is the possibility of an outside influence finally eliminated. But this is an aspect of the theory whose implications were only understood much later, as we will see in subsequent chapters.) In the practical case, Newton's approach starts from the orbit of a single point mass around a spherically symmetric central mass, and promises that the transition from this simple model to the actual motions of the Solar System will yield an increasingly accurate explanation of irregular motions of its members by their mutual perturbations.

Fulfilling this promise – to a reasonable degree of approximation – was not as straightforward as it may seem in retrospect. While Newton did provide a remarkably accurate account of the motions of the Solar System, he was unable to account for the orbit of the Moon and the anomalous motions of Jupiter and Saturn; these problems were only solved, respectively, by Clairaut in 1749 and Laplace in 1785 (see Wilson, 2002). Moreover, beyond celestial mechanics, important classes of motion eluded Newton's grasp entirely, especially fluid dynamics and rigid body motion. Careful historians have emphasized, therefore, that what we now think of as Newtonian mechanics, as a comprehensive mathematical theory of motion and force, was completed only by the combined efforts of Clairaut, Euler, D'Alembert, Lagrange, and Laplace. [As Truesdell has expressed it, "what physicists today call Newtonian mechanics has little direct relation to Newton's own work, but is rather a combination of Euler's mechanics with Lagrange's" (Truesdell, 1967, p. 252).][10] This work involved, moreover, a radical transformation in mathematical methods, from Newton's synthetic method to the analytic methods championed and developed by the Continental mathematicians – including, of course, Leibniz – without which the problems of rigid bodies and fluids, as well as the recalcitrant planetary motions, could hardly have yielded to Newton's laws. So the path from Newton's *Principia* to the science of "Newtonian mechanics" took nearly a century of difficult mathematical work. Nonetheless, there was little doubt (at least after Leibniz) that this work presupposed the Newtonian framework of space and time – that it must do so, because its fundamental task is to comprehend forces through the changes that they cause in states of motion.

It is significant that it was Euler who, in this time, most clearly articulated the presuppositions about space and time that lay behind all of this work in mechanics. For Euler was among the most sympathetic to Leibniz's metaphysical views: he was suspicious of action at a distance, and throughout his life he continued to hope for a workable vortex theory to replace the theory of attraction. At the same time, however, he understood that Leibniz's views of space and time could never be reconciled with the project of physical explanation in which he, along with the other great successors of both

Newton and Leibniz, was engaged. How did Euler come to this critical view of Leibniz, despite his sympathy for some of the central Leibnizian ideas? At least part of the explanation is evident: he combined elements of the Leibnizian metaphysics with a central element of the Newtonian *methodology*. Euler was willing to grant that matters on which Leibniz had taken a-priori positions – based on his metaphysical understanding of substance, causality, and "the nature of body" – were better seen as *empirical questions*. The merits of Newton's theory of attraction, in particular, must depend on whether the theory can deal with the unsolved problems of celestial motion. And the question of absolute and relative motion must depend on whether physics has the empirical means to distinguish them. What physics can tell us about motion, on empirical grounds, is our sole guide to what metaphysics can tell us about space and time. In other words, Euler saw the difference between the elements of Newton's theory that were, so to speak, idiosyncratically Newtonian – above all the idea that universal gravitation is the sole force at work in the Solar System – and those that represented the common basis of all work in mechanics as then understood, especially the laws of motion and their underlying framework of space and time. Thus he acknowledged the distinction between the physical hypotheses that one might prefer, pursue, and evaluate within the general framework of mechanics, and the conceptual framework without which such hypotheses could not even be intelligible.

It is true that, in the hands of Euler *et al.*, the treatment of motion and forces took on a fundamentally different form, moving from Newton's intuitive geometrical representation to the analytical representation, and the predominance of variational and conservation principles. But it is important not to allow this transformation to obscure the underlying continuity, as indeed Euler and his contemporaries clearly understood. For the mechanics developed in the eighteenth century, by mathematical methods that were largely Leibnizian in origin, is still Newtonian in its fundamental principle regarding space and time: that bodies, or systems of bodies, have a privileged way of evolving from a given state to a future state, and that this evolution embodies the idea of a privileged trajectory in space-time. In fact the most serious challenge to this general picture finally came, not from a relational theory of space and time, but from a completely unexpected direction: from the theory of the microstructure of matter, in which spatio-temporal order gives way to structures of another sort altogether, structures better characterized as algebraic or logical rather than spatio-temporal. The challenge, in other words, came not from "general relativity," in which space-time trajectories play the same kind of fundamental role that they play in the Newtonian theory, but from quantum mechanics, in which the very idea of a trajectory is called into question.

2.9 NEWTON'S ACCOMPLISHMENT

Did Newton succeed in proving that space, time, and motion are "absolute"? From any reasonable empiricist standpoint, he could not possibly have proven this. Whatever he might have shown could only be established in the context of Newton's laws of motion, and so, in order to establish any conception of space and time, he would have had to establish these very laws – a task that he himself recognized was impossible, since laws as we know them could always turn out to be "liable to exceptions" (Newton, 1726 [1999], p. 796), and his entire procedure could have to give way to some "truer" method of philosophizing (Newton, 1726 [1999], p. 383). As we have seen, however, he never attempted to prove any such thing. What he was in fact trying to prove, he did prove to a great extent: that his contemporaries, sharing as they did his fundamental dynamical assumptions, had no standing to criticize his conceptions of space and time, or to propose a more relativist view. For the mechanical philosophers, then, there were only three legitimate ways to resist Newton's dialectical argument. One would be to acknowledge it, but to insist upon its limitations: it doesn't prove, after all, everything that Newton had wished to prove. For it supports absolute time and absolute acceleration, but not absolute velocity and absolute space. Therefore it points to the need for a weaker structure than absolute space. But, again, no one in the seventeenth century was in any position to identify such a structure, and the unpublished remarks of Huygens on rotation are the closest approach to this problem until the late nineteenth century. A second way is to argue that, since the laws of motion do presuppose something like the Newtonian conception of space, time, and motion, they ought to be replaced by laws that don't presuppose any such thing. This alternative, too, was naturally beyond the conceptual horizons of the mechanical philosophers, and was first clearly expressed in the nineteenth century by Mach. The third way is, therefore, the only one truly available to someone like Leibniz: to maintain that no matter what spatio-temporal concepts might be required by physics, the authority to pronounce upon the true nature of space, time, and motion ultimately belongs to metaphysics, which can understand space and time on its own independent grounds. As we will see in the next chapter, even this possibility was not brought to fruition by Leibnizian metaphysics, and, indeed, its failure in this regard turned out to play a central role in the general downfall of Leibnizian metaphysics in the course of the eighteenth century.

To acknowledge the force of Newton's arguments against his contemporaries, then, is not to claim that they could finally settle the absolute–relational controversy. After all, no argument from classical mechanics could ever rule out the possibility that a "relational" theory of motion might be constructed that is empirically successful, even as successful as Newton's theory. But we can point to a lasting philosophical accomplishment of Newton's work. He showed that the philosophical understanding of space and time has to start, not from general philosophical principles, but from a critical analysis of what we presuppose in our observation and reasoning about the physics of motion. The eventual overthrow of Newton's theory was made possible by the further pursuit, in a different theoretical and empirical context, of the same kind of analysis.

NOTES

1. The most strident assertion of this view can be found in Reichenbach (1924).
2. In addition to Weyl (1918) and Eddington (1923), see, for example, Fock (1959), Synge (1960), Trautman (1965, 1966), Misner *et al.* (1973) or Ehlers (1973a, b).
3. Kant held that inertia is strictly the "inactivity" of matter, and that this inactivity alone cannot be the cause of resistance, for "nothing can resist a motion but the opposite motion of another, never the other's rest" (Kant, 1786 [1911], p. 551). See also Chapter 3, later.
4. For further discussion of the origin and philosophical significance of the concept of inertial frame, see DiSalle (1990, 2002d).
5. The four-dimensional affine space that expresses this same structure, "Newtonian space-time," is discussed in Stein (1967).
6. This interpretation of Mach's arguments is defended in DiSalle (2002c), to which the reader is referred for further discussion of the context of Mach's work, and its implications for twentieth-century philosophy of physics.
7. For an extended discussion of this question and its place within Newton's metaphysics in general, see Stein (2002).
8. This is a Kuhnian claim that I concede here for purposes of argument, but that is in fact quite unhistorical. While it is true that Galileo's standard of physical explanation is incommensurable with Aristotle's, it is certainly not a revolutionary departure from the prevailing standards of sixteenth-century mechanics. In fact the precise quantitative explanation of motion – especially projectile motion – was already an important preoccupation of the impetus theorists who preceded Galileo. So the change in scientific criteria is a more gradual one than the Kuhnian model of "paradigm shift" suggests.
9. It is worth pointing out that Kuhn noted something like this feature of Galileo's reasoning, in his discussion of some of Galileo's thought experiments. "In

short, if his sort of thought-experiment is to be effective, it must allow those who perform or study it to employ concepts in the same ways they have been employed before. Only if that condition is met can the thought-experiment confront its audience with unanticipated consequences of their normal conceptual operations" (Kuhn, 1977, p. 252). But evidently Kuhn did not consider this issue in the context in which it is considered here, as illustrating the possibility of non-circular arguments for new conceptual frameworks.

10. For further discussion of Newton's eighteenth-century successors, and their work on the completion of Newtonian celestial mechanics, see Wilson (2002). The present account is also heavily influenced by Smith (2003a, b).

CHAPTER 3

Empiricism and a priorism from Kant to Poincaré

Newton presented not only a theory of absolute space and time, but a philosophical approach to the analysis of space and time quite unlike anything
contemplated by his contemporaries. It cannot be viewed as a complete
philosophical account of space and time, however, because it treats space
and time solely from the perspective of classical mechanics – that is, as
concepts implicitly presupposed by the classical mechanical understanding of causality and force. A philosophically thorough treatment of the
problem would embrace, not only the implicit metaphysics of physics, but
the general epistemological problem of space and time and the ways in
which physics, and human knowledge generally, have some access to them.
In other words, the step beyond what Newton accomplished required an
attack on what later became known as "the problem of physical geometry."
The revolutionary development of space-time geometry in the twentieth
century, in both special and general relativity, is only the most spectacular
of the many far-reaching consequences of this philosophical effort.

As we saw, Newton's theory was forced into confrontation with the
most prominent general philosophical accounts of space and time, namely
those of Descartes and Leibniz. But its rejoinder to them was only that
those philosophical views could not be reconciled with their own views
of physics. Undoubtedly this was a compelling argument as far as it goes,
one which neither Descartes nor Leibniz was in a position to answer on
its own terms. On the assumption that physics describes the real world –
more precisely, that the physical conception of force as contained in the
laws of motion truly captures the metaphysical nature of causality – that
the physical account of forces captures the true nature of causal interaction
among things – there could be no question of the authority of physics to
speak to the fundamental nature of space and time, and no question of the
force of Newton's arguments. But if metaphysics had any claim to deeper
knowledge than physics, penetrating to the inner nature of things beyond
the ken of empirical science, then arguments like Newton's would be easy

to evade. One could simply retreat from the sensible to the intelligible world: "absolute" space, time, and motion, as understood in Newtonian physics, could be viewed as mere phenomena with no basis in the world of intelligible things and their intelligible causal relations.

The first requirement for a complete empiricist picture, therefore, was a critique of the pretensions of metaphysics. Newton, as we saw, took some essential steps in this direction, criticizing the mechanists' narrow metaphysical conception of causality. But it was Kant who, over the course of his career, subjected both Leibnizian metaphysics and Newtonian physics to the most systematic critical analysis. He argued that the former offered only an illusory promise of transcendent knowledge, while the latter offered a genuine metaphysics of nature, in which space and time were the basis for an objective understanding of force and motion. The Leibnizian metaphysics, in contrast, involved a confused conception of transcendent knowledge, an illusion that force and motion could be understood as purely metaphysical concepts independently of their representation in space and time. The only intelligible conceptual basis for a genuine metaphysics of force and motion, then, lay in Newtonian physics itself, in the representation of force as determined by spatial and temporal relations among masses. More generally, Kant's account of space – whatever its defects – showed that our knowledge of space is inseparable from our means of representing it to ourselves in experience. That is why this chapter on the development of physical geometry, a nineteenth-century movement that represents the overthrow of Kantianism, must begin with Kant himself.

3.1 A NEW APPROACH TO THE METAPHYSICS OF NATURE

The history of Kant's philosophy of science, and the role it played in the development of the critical philosophy, have been well documented by others. How it relates to the themes of this book, however, and to the ongoing concerns of the philosophy of space and time, deserves some further discussion. Two questions are particularly worth considering. How did Kant understand the role of absolute space in Newtonian physics? And how was this understanding, in turn, connected with his general project for the reform of metaphysics?

As Friedman has convincingly argued, Kant's analysis of Newtonian physics – throughout his career, but especially in the critical period – was remarkably insightful, even "a model of fruitful philosophical engagement with the sciences" (Friedman, 1992, p. xii). His insights concern, not only the philosophical implications of Newton's theory, but also its philosophical foundations, that is, the relation between its principles and the principles of metaphysics. Now, this second issue might be regarded in either of two ways. On the one hand, one might think of physics and metaphysics as competing sources of claims about the world: as the example of Newton and Leibniz clearly shows, both disciplines make claims – about the nature of space and time, motion and force, substance and causality – that may be fundamentally at odds. As we saw, the ultimate source of philosophical dispute may be the disagreement about which set of principles has the better claim to truth. On the other hand, one might think that physics requires a foundation in metaphysics; in that case, its principles have to be seen as superficial or uncertain, or both, until they can be shown to be *derivable* from deeper metaphysical principles. The two possibilities are evidently not quite mutually exclusive, but the second one, at least, entertains the possibility that the principles of physics may be taken at face value – even if physics itself is in no position to comprehend their deeper significance, or the sources from which they arise. That is to say, the second view acknowledges a kind of independence for physics, even a right to establish its own results by its own methods, but reserves for metaphysics the understanding of *why* these principles are true. The first view is best exemplified by Descartes and Leibniz, who felt free to reject ideas of Galileo and Newton that had no other basis than empirical evidence; the second is exemplified by Euler, who held that the empirically established principles of Newtonian physics were principles with which metaphysics must come to terms. It was also the view of the early (pre-critical) Kant, first in his Leibnizian–Wolffian phase, and then in his gradual turning away from the Leibnizian–Wolffian metaphysics; this is hardly surprising, since Kant's turn was directly and crucially influenced by Euler, and arose from his efforts to reconcile the Leibnizian tradition with Newtonian physics.

There is a third alternative, however, not so easily characterized, but absolutely decisive for understanding the progress of Kant's philosophy, his grasp of the Newtonian revolution, and that revolution in itself. It concerns not so much the metaphysical principles that may be seen as grounds or consequences of physical principles, but, rather, the nature of the *right* that physics can claim to address metaphysical issues at all. It concerns, that is, the question of what claim the mathematical–physical picture of the universe has to genuine intelligibility. It is a view of physics, in its Newtonian form at least, as being in dialectical engagement with the metaphysical tradition. Rather than being a source of alternative views from outside metaphysics – to be refuted, absorbed, or derived from metaphysical first principles – Newtonian physics is in itself a philosophical critique of metaphysics as traditionally practiced. This general viewpoint was already suggested by Newton himself, as we saw in the previous chapter. But it was Kant who developed and applied it in detail, and who made it the means of transforming philosophy as definitively as Newton had transformed physics.

The mechanical philosophy, as part of the tradition that Kant referred to as "dogmatic" metaphysics, claimed to present, at least, an intelligible picture of the Universe and the fundamental principles that govern it; it claimed to reduce every natural process to characteristic kinds of entity and interaction that were clearly understandable in mechanical terms. Thus the mechanical philosophy claimed to have created a standpoint from which the intelligibility of physics in general could be judged. The rejection of universal gravitation, by philosophers in this tradition, is only the most familiar example of this general claim. It was a conception of intelligibility that Kant himself had shared in his earlier works, and that lies behind his pre-critical notion of an "intelligible world." How Kant came to abandon this notion, and to reconsider the distinction between the noumenal and phenomenal worlds in general, is a familiar aspect of the history of the critical philosophy. More recently, the role that Kant's grasp of Newtonian physics plays in this history has been explained by Friedman (1992). I would only suggest that we view this development in a slightly different perspective, one that promises some further insight into the nature of Kant's break with tradition, as well as of the lasting relevance of his work. Kant's rejection of the mechanists' metaphysics was more than a side-effect of his general rejection of dogmatism. Yet is not enough to say, instead, that Newtonian physics provided intellectual stimulus and telling examples for that transformation in Kant's thought. To understand Kant's development completely, we have to understand how, in his view, the Newtonian revolution in natural philosophy had succeeded, and what was the nature of its success. For Kant, it was more than a scientific revolution whose implications changed metaphysics irrevocably. The Newtonian revolution was itself a revolution in metaphysics.

If this point has been hard to see since the later twentieth century, it is probably because of the prevailing interest in the question of scientific rationality, especially the rationality of scientific revolutions. At least since the major works of Popper and Kuhn, debates within the philosophy of science have focused on how, or whether, scientists change their theoretical convictions by some rational process. On the assumptions of such debates, all of the epistemic authority of science depends upon our ability to exhibit a scientific method for theory-choice that is both exemplary of rationality, and true to the historical and present practice of science. Unless we can do this, we will be unable to distinguish science from any other human "belief system." For Kant, however, the problem of rationality was not a pressing one. This is not because he took the rationality of science for granted without question; rather, it was because he saw the question of the epistemic authority of science in altogether different terms. For the question that he was trying to answer, rationality was completely insufficient.

The preoccupation with rationality was as much a part of Kant's context as it is of ours, though of course the problem was understood quite differently. For the Leibnizian tradition in metaphysics, in which Kant himself had been schooled, the grounds for rational belief in the principles of physics was a central question. The answer was supposed to come from a "metaphysics of nature," or a foundation of accepted metaphysical principles on which physics could be built; the rational grounds for belief in physics was to be that its principles could be logically derived from this metaphysical foundation. Leibniz's distinction between the "kingdom of final causes" and the "kingdom of efficient causes" implied that it was the former whose principles could be certainly known, derived from the laws of non-contradiction and sufficient reason; there was accordingly just as much truth in the laws of efficient causation – that is, the laws of motion as understood by the mechanical philosophy – as could accrue to them from their basis in the laws of metaphysics. What Kant called the dispute between "the metaphysicians" and "the geometers" concerned, in large part, whether the methods of mathematical physics, as they were then beginning to be understood, could justify belief in principles that went against what metaphysics claimed to know on deeper rational grounds.

Kant eventually realized, however, that despite the claim to a rational foundation, metaphysics had not produced any principles that could command the kind of universal assent enjoyed by the laws of physics. Metaphysics looked to mathematics as a model of epistemic certainty, but generally assumed that the certainty of mathematics was secured by its inherent rationalism, as embodied in its deductive method. Kant saw, however, that metaphysics already had – quite literally – more rationality than it knew what to do with; he saw that mathematical physics, in contrast, had something in addition to rationality that made it a source of clear a-priori principles instead of endless, aimless disputes. It was wrongheaded to expect to achieve the success of the exact sciences, merely by imitating their deductive structure. For metaphysics had shown, throughout its history, that it could be completely rational without ceasing to be completely subjective as well. What was lacking, then, was a way to be sure that its starting point was not arbitrary. This sort of assurance is precisely what distinguished the exact sciences. Mathematics, and mathematical physics, have no arbitrary elements because their fundamental concepts have not been in dispute: unlike those of metaphysics, the fundamental concepts of the mathematical sciences are constructed under rigid and objective constraints, constraints imposed by the nature of sensible intuition (Kant, 1764 [1911], pp. 276–8).

If this is true, however, it does not make sense to say that these concepts have, or need to have, a foundation in some prior privileged discipline, such as "the metaphysics of nature" or even "transcendental philosophy." When Kant criticizes physics for its lack of a genuine foundation – when he criticizes physicists for merely "postulating" laws without bothering to seek their a-priori sources – it is easy to get the misleading impression that a deduction from deeper premises is in order. Then the physical laws' relation to metaphysics would be like the relation of, say, Kepler's ellipse law to the principles of physics: we might demand to know, not merely that they are true, but why they are true, and the answer we might demand would be an explanation of the fundamental metaphysical principle from which they follow. This was the attitude underlying Leibniz's program to explain the laws of the "kingdom of efficient causes" – the world of interacting bodies as described by physics – by deriving them from the laws of "the kingdom of final causes," or the world of monads. But such a view, however natural it might have seemed to Kant in his pre-critical "dogmatic" period, could hardly be coherent with Kant's transcendental philosophy. The power of physics to *construct* an intelligible conception of the world means that physics is not a *consequence* of the metaphysics of nature. Quite simply, it *is* the metaphysics of nature. The metaphysical concepts that occur in physics – body, force, motion, space, time – become intelligible to us precisely, and only, as they are constructed by physics itself; physics provides us with the only intelligible notions we have on these matters. In other words, the mathematical sciences do describe an intelligible world, and, because of their empirical method of constructing fundamental concepts, that world is the sensible world itself.

3.2 KANT'S TURN FROM LEIBNIZ TO NEWTON

Kant's conception of physics and its true relation to metaphysics, as just sketched, defines the setting in which we must interpret his ideas about absolute space and time. As was noted in the previous chapter, Kant had started from a Leibnizian view of the world as constituted of monads, and consequently a relationalist view of space; he was moved in the direction of Newton's view largely by his reading of Euler.[1] Evidently Kant was impressed by the argument cited in Chapter 2, that dynamics must assume certain aspects of space and time – above all, the idea of a privileged state of uniform motion – that cannot be squared with Leibniz's relationalism. From Euler, Kant also learned of the apparent contradiction within Leibniz's picture: if space is reducible to spatial relations, and motion therefore to change of relative position, what sense can be made of Leibniz's dynamical notion of force? How can the idea of force as a genuine metaphysical quantity be reconciled with the relativity of motion? Leibniz himself occasionally juxtaposed these notions, apparently unaware of the conflict that others saw quite clearly. To Huygens, for example, he wrote, "But you will not deny, I think, that each [body in a group of interacting bodies] does truly have a certain degree of motion, or, if you wish, of force, in spite of the equivalence of these hypotheses about their motion" (Leibniz, 1694, p. 184). Even in his correspondence with Clarke, in the course of his most detailed arguments for the relativity of motion, Leibniz acknowledged that "there is a difference between an absolute true motion of a body, and a simple relative change of its situation with respect to another body. For when the immediate cause of the change is in the body, that body is truly in motion" (Leibniz, 1716, p. 404). As Euler put it, "Therefore I am least afraid of those philosophers who reduce everything to relations, since they themselves attribute so much to motion that they regard moving force as something substantial" (Euler, 1765, 2:79).

Yet there could be a proper Leibnizian reply to such objections, even if Leibniz did not appreciate the force of the challenge well enough to make it. He might answer, simply, that they represent a confusion of physics with the metaphysics of nature. Force, as Leibniz understood it, is a quantity with a metaphysical foundation in the internal state of a monad; it is an expression of monadic appetition, the "striving" of the monad toward a future state of existence. The metaphysical reality of such quantity must be understood completely independently of phenomenal space and time; it represents that aspect of physics that is "more than geometry can determine," because it concerns the inner activity of substances, rather than their phenomenal interactions. The latter, after all, are merely apparent in Leibniz's system, merely a confused human way of seeing the pre-established harmony that brings their actions into correspondence with one another, without any genuine mutual influence. This concept of force thus provides the phenomenal world of interacting bodies with a foundation in the intelligible world of "windowless" substances. If that is true, it cannot make sense to criticize or to revise the metaphysical foundation because of what might appear to be the case in its phenomenal reflection. Nor can the existence of an intensive, metaphysical quantity of force in the metaphysical world justify granting absolute reality to motion, space, or time as understood on the purely phenomenal level of empirical science.

Even if this argument was never put forward explicitly in these terms, it helps us to understand why Kant, when he took up the subject of space himself, was not content with Euler's defense of absolute space. Instead, in "Concerning the ultimate ground of the differentiation of directions in space" (Kant, 1768 [1911]), he was eager to confront the Leibnizian position on more general philosophical grounds. He sought to show that Leibniz's relationalism was incompatible not merely with physics, but also with a more general feature of space that can be exhibited without any reference to the special assumptions of physics. Thus he would "provide, not engineers, as Euler had in mind, but geometers themselves with a convincing ground, with the evidence to which they are accustomed, for claiming the actuality of their absolute space" (Kant, 1768 [1911], p. 378). This would be an appropriate sort of argument to make against a disciple of Leibniz, since Leibniz held the certainty of geometry to be beyond question, and even claimed that its principles could ultimately be deduced from logical identities. If the argument were to succeed, it would not only vindicate "the geometers" – including the followers of Newton's "mathematical principles of natural philosophy" – against "the German philosophers" who followed Leibniz. It would also help to establish just the sort of connection between the intelligible and the sensible – between fundamental truths of reason and the space of our immediate experience – on which a true metaphysical account of space could be built.

The 1768 paper presents Kant's well-known argument from incongruent counterparts: pairs such as the right and left hand represent equal objects from the point of view of their (Leibnizian) internal spatial relations, yet they are incongruent in the sense that they cannot be superimposed on one another; the fundamental difference between them concerns not the spatial relations among their parts, but their relations to space itself. This argument is undoubtedly the most commented-on of all Kant's arguments about absolute space,[2] and it is not difficult to see why. More than any other discussion of Kant's, it has the general form of a gambit in the absolute–relational debate: it attempts to exhibit a known phenomenon that the relationalist view cannot account for, and whose explanation appears to require the existence of absolute space. But, whatever its intrinsic merits, there are two good reasons not to take the 1768 argument seriously as a discussion of absolute space, at least as far as the themes of this book are concerned. One is that, in his critical phase, Kant would no longer admit that either the Newtonian or the Leibnizian side could possibly be correct; both are founded in dogmatic metaphysics. Even in 1770, in the Inaugural Dissertation, Kant would no longer wish to defend the thesis that "Absolute space, independently of the existence of all matter and as itself the first ground of the possibility of the composition of all matter, has a reality of its own" (Kant, 1768 [1911], 2:378). From 1770, Kant came to think of space and time as the forms of inner and outer intuition. There was no question of the absolute reality of space and time once they were acknowledged to be "subjective and ideal," aspects of our own sensibility; the Newtonians and the Leibnizians, therefore, shared the delusion that space and time belonged to things in themselves – in the Newtonian case, as themselves real things, and in the Leibnizian case, as founded in relations of real things. That is to say, they were both working within the "dogmatic" tradition from which Kant had finally escaped. To the extent that Kant discusses incongruent counterparts at all in his critical period, he does not associate them with the problem of absolute space, but with the problem of spatial intuition, in particular with the claim that there is an intuitive representation of space that cannot be reduced to concepts (e.g. Kant, 1783, section 13). In the setting of Kant's critical project, the question cannot be whether a particular metaphysical view of space can explain particular phenomena; the question is, rather, what assumptions about space are conditions of the possibility of experience.

The other reason to discount incongruent counterparts is less obvious, perhaps, but more important. Their existence is simply irrelevant to the question of absolute space: their existence is a question concerning only the structure of three-dimensional space, whereas absolute space is, as we have seen, a theory of space-time. Like Leibnizian arguments about spatial reflections and translations, then, Kant's arguments have nothing to say about the way in which space is connected through time. We could know certainly that the distinction between left and right is an inherent feature of the natural world, without getting any insight at all into whether any of the claims in Newton's Scholium are true. In short, the discussion of incongruent counterparts is the part of Kant's writing that most directly touches on the "absolute–relational" debate in its traditional sense. At the same time, however, and for the same reason, it is the least insightful of Kant's discussions of absolute space – the one furthest removed from the essential issues concerning Newton's concept of absolute space, namely, the role it is supposed to play in physics, and its interconnections with the concepts of mass and force. Such an assessment is, of course, obvious from the point of view of this book, but it was Kant's assessment as well: his most advanced discussion of absolute space, in the *Metaphysische Anfangsgründe der Naturwissenschaft* (1786 [1911]), focuses on its relation to the concept of absolute motion, and on the relation of both concepts to the application of Newton's laws. In contrast to the 1768 argument, the later discussion contains insights into the implications of Newton's theory, both for physics and for metaphysics, that are only beginning to be noticed.[3] It reflects Kant's later awareness that, after all, physics does have an answer to Leibniz's metaphysical objections: the metaphysical concepts underlying the sensible world first become intelligible, themselves, in the framework of Newtonian physics.

3.3 KANT, LEIBNIZ, AND THE CONCEPTUAL FOUNDATIONS OF SCIENCE

Kant's concern about the arbitrariness of metaphysical concepts began well before his critical turn, at least as early as the "Prize Essay" of 1764. The problem, he thought, lay in the method by which those concepts are defined. Metaphysics must proceed by the analysis of concepts about which people have some vague associations, but no precise definition. Mathematical definitions, by contrast, are synthetic, arrived at simply by the construction of the object, either in imagination or on paper; they are therefore constrained by the laws of our own sensible intuition (Kant, 1764 [1911], pp. 276–7). As a result, there can be no doubt about whether a mathematical definition captures what it is intended to capture, or about which objects fall under it. But metaphysical definitions are always subject to such a doubt, for there is no established way to determine whether such a definition contains everything that is proper to the concept to be defined. What considerations, after all, could assure us that a metaphysical definition captures some pre-existing meaning, rather than merely assigning one arbitrarily? Therefore the philosophical analysis of a concept such as God or substance can never come to a certain or universally satisfying end. If mathematics mistakenly tries to define a concept by analysis, no harm is done, since mathematical reasoning will require that the concept be constructed in any case, and it is the constructed concept that will matter to the reasoning. But if metaphysics tries to define by synthesis, in the absence of the constraints imposed by sensible intuition, the likely result is a concept that is nothing more than an arbitrary invention – like the Leibnizian monad (Kant, 1764 [1911], p. 77). A philosophical system built on such concepts inevitably has the character of a mere hypothesis. Leibniz himself, on occasion, went so far as to admit the hypothetical character of his system,[4] though he generally claimed to have derived it from first principles. As far as Kant was concerned, however, Newtonian science had dispensed with hypothetical foundations, and metaphysics could not claim universal assent until it had done likewise.

It is true that Kant's view of mathematical concepts was, like other central parts of the critical philosophy, undermined by the progress of the nineteenth century. The rigorization of analysis, beginning with the work of Bolzano (1817), provided a clear example of what Kant had held to be impossible: the construction of mathematical concepts, especially of infinity and continuity, "by means of mere concepts" and without appeal to intuition. His belief that intuition was indispensable, as many commentators have pointed out, only reflected the limitations of logic as he understood it. But this development does not undermine his critique of the Leibnizian tradition. On the contrary, it reveals how crucially important that critique was for the further development of mathematics. If Kant erred in thinking that mathematical reasoning *must* appeal to intuitive constructions, he was correct in thinking that mathematicians did *in fact* rely on intuition – even mathematicians who, like Leibniz, imagined that their grasp of mathematics, and the foundation of its truth, were purely intellectual. Nineteenth-century philosophers could not have purged mathematical reasoning of intuitive steps, surely, had they remained under the Leibnizian illusion that there were none.

Because this illusion was so pervasive, and so much rested upon it, the consequences of its exposure were profound. As Kant pointed out, Leibniz's view could provide no basis for the truth of geometry, or of any part of mathematics. On the one hand, as Kant noted, geometrical proof requires appeal to construction and, therefore, the validity of Euclidean theorems is in doubt without some use of intuition. On the other hand, the very existence of the objects of geometry has no other basis than intuition. To the first point, Leibniz could claim to have an answer in the "universal characteristic": the need for constructive proof, in that case, would appear to be a sign of the inadequacy of our tools for the logical organization of knowledge. Of course Leibniz never succeeded in providing an adequate tool, but he could be excused (and was eventually vindicated) for believing that such a thing was possible. To the second point, however, it is not clear what sort of answer Leibniz could have been able to give. If geometry concerned the relations among "things in themselves," then, as Kant frequently emphasized, the truth of its principles would always be incapable of verification, and therefore in doubt (A40/B56); the idea that geometry concerns merely empirical relations was, evidently, not an available alternative for Leibniz. Furthermore, even if the "universal characteristic" could capture the logical form of geometrical arguments, it could never justify the claim that geometrical statements are necessarily true of the objects to which they refer – a claim that Leibniz believed as surely as Kant did. Leibniz rested the truth of geometry, as of every science that is necessary and universal, on the fact that its principles could ultimately be reduced to logical identities. Therefore he did not see what Kant saw, namely, that even if such a reduction were possible, it would come at the cost of geometry's reference to any real object. On this point it was Kant's view that was vindicated in the nineteenth century, indirectly, since the logical apparatus that made it possible to understand geometry as a logical structure, and different geometries as intertranslatable structures, involved the separation of the structure from content (see Nagel, 1939).

3.4 KANT ON ABSOLUTE SPACE[5]

To understand Kant's rejection of the Leibnizian view of space, then, it is insufficient – and possibly counterproductive – to connect it with his efforts in the traditional absolute–relational debate. Instead, we have to see it in connection with the more general problems of philosophy, and so to grasp its connection with Kant's more general critique of the Leibnizian approach to metaphysics. His mature concern was not to establish one of two opposing metaphysical positions, but, rather, to understand how metaphysics in general can know what it is talking about – more precisely, how metaphysics might, like the mathematical sciences, define its fundamental concepts in such a way that they are no longer matters of subjective opinion and controversy, but acknowledged as universal and necessary. This represents a completely changed view of the nature and purpose of a philosophical "system." Where Leibniz's system was quite clearly a hypothesis meant to "make sense of" the world, providing it not only with an origin and structure but also with meaning and purpose, Kant's metaphysical system is meant only to capture the concepts without which it would be impossible to think of a world at all. The former, perhaps, best corresponds to the familiar use of the phrase "philosophical system," but as Kant realized – and this is arguably the fundamental distinction between pre- and post-Kantian philosophy, even if it is not always respected – such systems are irretrievably subjective or even arbitrary. This fundamental change in philosophy, moreover, mirrors the change in science from the mechanical philosophy to the Newtonian "mathematical" philosophy. The principles of the mechanical philosophy, though represented as the only hope for an intelligible picture of nature, are in the end only hypotheses about the ultimate nature of physical interaction, and therefore they do have something of the arbitrariness of philosophical hypotheses. But Newton's mathematical principles represent (for the eighteenth century, at least) the general conditions under which any causal interaction can be comprehended under natural laws. Kant's analysis of absolute space, accordingly, is an effort to clarify its place within the system of Newtonian principles. Though the effort did not completely succeed, the analysis of Newtonian physics placed its relation to metaphysics, and the nature of metaphysics in general, in an exceptionally revealing light.

To spell out Kant's account of absolute space is not without difficulty, and commentators have differed widely, not merely on the usual subtleties of interpretation – as they do regarding all aspects of Kant's philosophy – but even on what, if anything, the account actually says. Earman, for example, finds Kant's position inconsistent, and seems to despair of making any sense of it (Earman, 1989, pp. 76–8). Undoubtedly Kant, like most philosophers before the late nineteenth century, had some trouble reconciling the only reasonable position on absolute space – that it is superfluous to Newtonian physics – with the only reasonable position on rotation and acceleration – that they are essential parts of the theory. As we have seen, in spite of the fact that Kant's transcendental idealism compelled him to reject Newton's view as surely as Leibniz's, he found it impossible to dispense with absolute space. Before condemning the inconsistency, however, it is worth recalling that Kant's transcendental idealism was also an empirical realism and that "this [empirical] reality of space and time leaves the certainty of our experiential knowledge untouched: for we are just as certain of it, whether these forms necessarily depend on the things themselves or only on our intuition of them" (A39/B56). It might be said, then, that instead of creating an inherent difficulty for his assessment of absolute space, Kant's transcendental idealism sets aside the ontological controversy, and leaves him free to consider absolute space only insofar as it has a certain function in Newton's dynamical theory. As a result, his account of absolute space brings out some aspects of it that were not made explicit enough by Newton, and that tend to be obscured within the terms of the "absolute–relational" controversy.

On the most common way of approaching the subject, in the terms of the absolute–relational debate, the role of absolute space is summed up in the claim that "absolute motion is a species of relative motion" (see Sklar, 1977, p. 229; Earman, 1989, p. 13), i.e., that absolute motion means motion "relative to" space rather than to the material environment; on that interpretation, the theory of absolute space asserts that we cannot understand the dynamical effects of motion if we refer it to visible bodies, and so we explain the effects by referring motion to absolute space. In these terms, absolute space comes across as an extremely dubious notion. Indeed, part of the puzzlement of the relationalist is provoked by the very notion that something unobservable has any possible value as a reference frame. For the same reason, it would seem that the only possible defense of absolute space is that it represents a kind of "inference to the best explanation" (see Earman, 1989, pp. 63–4), an unobservable "theoretical entity" postulated for its potential explanatory power. As we already noted, this was certainly not the position of Newton, whose arguments for absolute motion and absolute space were not hypothetico-deductive. But Kant went much further, and analyzed more explicitly the special position that absolute space occupies in relation to the ontology of Newtonian physics. Though not without its own genuine obscurities, his account does help us to see the matter in the proper light.

From Kant's perspective, it is obvious from the start that absolute space, since it is no object of experience, cannot serve as a relative space. Instead, absolute space has the function of a rule for considering the interactions among bodies.

Absolute space is therefore necessary, not as a concept of an actual object, but as an idea that is supposed to serve as a rule for considering all motion within it as merely relative, and all motion and rest must be reduced to absolute space, if the appearance of the same is to to be transformed into a determinate concept of experience (which unites all appearances). (Kant, 1786 [1911], p. 560)

That is, the motions of bodies are not imagined to be "referred to" absolute space; rather, they are subjected to a dynamical analysis, in whatever reference frame we might find practical, and thereby "reduced to" absolute space. Therefore it is no objection to absolute space that we cannot use it directly as a measure of velocity. Its place in physics depends only on whether we can attach some physical meaning to the concept of absolute velocity, by exhibiting some dynamical reasoning by which it can be said to be known. And while this cannot be done for the velocity of a body in empty space, it does seem to be possible for the velocities of a pair of interacting bodies. For, while their relative velocity may be grasped from any number of equivalent points of view, there is a privileged point of view from which their velocities are truly equal and opposite, namely the frame of reference in which their center of mass is at rest: in that case their two velocities "destroy one another in absolute space," and the interaction between them is thus "reduced to absolute space" (Kant, 1786 [1911], p. 545). Thus "there is no absolute motion" in the sense that a body's motion has no objective meaning in relation to space alone. But a body does have an absolute power to communicate motion to another body, when the two interact in accord with Newton's third law, and this constitutes its true state of motion.

This analysis has a limitation of which Kant was very well aware. The "absolute" velocities determined in this way are only the bodies' velocities relative to their center of mass frame. That frame itself, however, must be considered as moving relative to other centers of mass. Therefore the analysis may be said to be a reduction, not to absolute space, but to a particular privileged frame of reference. It might appear that in acknowledging this, Kant has implicitly acknowledged the equivalence of all inertial frames, and thereby avoided the error in Newton's conception of absolute space. But this is not quite true. If no actual center of mass frame is privileged over all others as being absolutely at rest, it is not because in Kant's account they are all equivalent descriptions of motion. Rather, it is because they are all necessarily *partial*; each system of masses must be understood as included within a still larger system of masses, as the Earth and Moon in the Solar System, the Solar System in the Galaxy, and so on. Its motion is deemed "reduced to absolute space" because the analysis of motion in such frames is supposed to lead, in the ideal limit, to a single frame in which all interactions in the Universe are to be comprehended.

We saw that Newton had expressed his awareness of this difficulty in his application of Corollary VI to the Solar System: if the entire system could be accelerated by external forces without our knowing it, then the accelerations of all the bodies in the system, as determined by their mutual interactions, could hardly be known to be their true accelerations. But Kant saw the significance of this fact to a degree that was not surpassed until the advent of general relativity.[6] In other words, Kant noted something about Newton's celestial mechanics that was not fully appreciated until it was considered by Einstein: the idea of true acceleration is, practically speaking, nearly as questionable as the idea of true velocity.[7] The only difference is that we actually can conceive of that ideal limit of a universal center of mass frame, without violating the relativity principle that is inherent in Newtonian mechanics. Yet Kant himself did not fully embrace the Newtonian relativity principle, since, in his view, the ideal limit is one in which the absolute velocities are known, not merely one in which the true accelerations are known. The most we can say is that absolute space, for Kant, has the same abstract function as an inertial frame. It is understood, as it were, "functionally," as the frame in which the true quantity of motion is known, at least provisionally. The reason for understanding such a frame as absolute space has something to do with the role played by intuition: the true frame of reference is one in which the true motions can be constructed in intuition, or represented to intuition. That is, if such a frame really could be identified, a body's observed velocity relative to it would necessarily constitute its true velocity.

For Kant, this aspect of absolute space reveals the essential inadequacy of the Leibnizian point of view. It is not merely, as Euler had already suggested, that Leibniz's account of moving forces attributes a kind of reality to motion that seems inconsistent with the relativity of motion and space. For that would be merely a kind of hypothetico-deductive difficulty, in the sense referred to earlier. It would be a difficulty, in other words, of reconciling a metaphysical hypothesis with the actual content of physics. The more serious difficulty is within Leibniz's metaphysics itself. The very concept of force that Leibniz employs has no meaning – or, more precisely, it has no meaning but what it acquires by being constructed within the framework of absolute space. A Leibnizian might still try to defend the idea that the metaphysical concept of force is independent of any such connection with the physical concept. But that is tantamount to abandoning a crucial part of the broader Leibnizian program: namely, the construction of a metaphysical foundation for physics. For that part of the program, at least, depended on articulating a connection between the metaphysical concept of force and the physical concept. The physical concept, however, is explicated as the power to generate motion; the moving force of one object is no other than its power to change the motion of another. Thus Leibniz, though he refers to it as a kind of intensive magnitude, has no other account of what moving force is than as what determines the action of one body on another. Therefore Kant points out that it has only one clear representation as a quantity: namely, as the power to generate a given velocity in a given space. It has an objective representation only when this quantity is constructed as part of an interaction, satisfying Newton's third law and therefore, in Kant's sense, "reduced to absolute space." In other words, motion belongs to the group of fundamental concepts for which Leibniz had thought to provide a purely metaphysical understanding, but whose definition is possible only within the framework of space and time. Leibniz's wish to see it as a kind of intensive magnitude is ultimately illusory.[8] The flaw in Cartesian physics, acccording to Leibniz, was the attempt to reduce body to extension alone; the Cartesians thus ignored the connection between motion and force, which is something "more than geometry can determine." But, as Kant's analysis shows, force in Leibniz's sense is something that cannot be comprehended, or even represented, independently of geometry.

From Kant's perspective, the reason for emphasizing these confusions on Leibniz's part is not simply to score philosophical points at his expense. Rather, it is to emphasize the limitations of Leibnizian metaphysics, in particular its incompatibility with the newly developed science of mechanics. And Kant's point goes far beyond the one made by Euler, and shared by himself in his pre-critical period – that is, that the established laws of physics illuminate the nature of space and time more than the doubtful conjectures of metaphysics. Kant is arguing that physics, in its Newtonian form, has arrived at clear concepts of space, time, motion, and force, where metaphysics has only been mired in confusion. It is not an attempt to judge between competing metaphysical hypotheses, according to their conformity with the facts of physics. It is, rather, the recognition of physics in its transcendental role, as the source of constructive definitions for metaphysical concepts.

This last point sheds some light on Kant's objections to Newton, and his complaint about physicists' willingness merely to postulate fundamental laws, without taking any interest in the laws' "a priori sources" (Kant, 1786 [1911], p. 472). As we have seen, it is not true that Newton was content to regard absolute space, and the laws of physics generally, as mere hypotheses. In fact the arguments that he presented may be characterized as transcendental arguments of a sort. The concept of absolute motion, he argued, was implicitly assumed in Cartesian physics, and was in a sense a "condition of the possibility" of Cartesian reasoning about the motions of the Solar System and their physical causes. But this is a kind of "relatively" transcendental argument: the concept is necessary, relative to a certain well-established practice of scientific reasoning about a certain kind of phenomenon. Kant, instead, argues that the Newtonian concept, and the laws of motion, are "absolutely" transcendental. They are the *only* basis on which the concept of causality can be applied to the Universe at large. They are the only basis, indeed, on which the phenomena of the heavens can be grasped as something more than mere appearances – as the appearances of genuine physical objects that stand in objective geometrical and causal relations.

The foregoing helps us to understand why Kant would have regarded universal gravitation in such a different way from Newton, as something that is "essential to matter" in just the sense that Newton always resisted (see Friedman, 1990). To Newton, the laws of motion provided the necessary and sufficient framework for understanding the interactions of bodies in space and time; gravitation was in no sense part of this framework, but was a kind of interaction that could be "deduced from the phenomena," as soon as the phenomena are interpreted within that framework. Questions about its properties, such as whether it acts immediately at a distance or by propagation through some medium, were open empirical questions, just as gravity itself was an answer to an empirical question about the nature of the interplanetary force. But in Kant's view, gravitation played a much more fundamental, even a transcendental role, as something that is indispensable to our understanding of matter and motion in general.[9] It is only by means of universal gravitation, considered as an immediate action at a distance, that Newton is able to consider the planetary system as something approximating an inertial frame, for it is only through the gravitational interaction that any estimate of the masses involved – and hence of the center of mass – is possible. But this is only an instance of a more general circumstance, namely, that it is only through gravitation that the principle of causality can be applied to the celestial bodies. Kant had some reason, then, to think that gravitation was something more than merely an empirical fact discovered within the Newtonian framework, and in fact was more closely tied to the basic principles of the framework than Newton realized. Just how closely tied, of course, is something that would eventually become clear with general relativity, through the understanding that gravity could not be separated from inertia.

3.5 HELMHOLTZ AND THE EMPIRICIST CRITIQUE OF KANT

It seems obvious that Kant's view of space as the a-priori form of outer intuition was overturned, in the course of the nineteenth century, by empiricist views of space and geometry. It should be clear from the foregoing discussion, however, that this is a somewhat simplistic and misleading picture. The very distinguishing feature of Kant's view, in regard to the earlier tradition in metaphysics, is its empiricism; his chief objection to the Leibnizian view of mathematics and science was its claim to knowledge of an intelligible world, through concepts whose only genuine content is that which they acquire from sensible intuition. Kant's empiricism in this regard was fairly radical, in fact, given the pervasiveness, even among empiricist philosophers such as Hume, of the assumption that mathematics is purely formal. It would be more illuminating, therefore, to distinguish among some empiricist tendencies that challenged Kant's peculiar form of empiricism, and its conception of the synthetic a priori. The most straightforward challenge came from an inductivist standpoint like that of Mill (1843):[10] the principles of geometry are held to be inductive generalizations from the practice, over many generations of human beings, of ordinary spatial measurement. The axioms of geometry, on this view, acquire their aura of certainty from the innumerable instances in which they have been confirmed by experience. It is quite understandable that after the twentieth century, when general relativity gave prominence to the idea of measuring the curvature of space and testing Euclidean geometry, this inductivist account would appear to have been the most important breakthrough.

Yet the inductivist account was not the view that decisively defeated Kant. The most important reason was, simply, that it could not account for the features of geometry that were central to Kant's theory. Given the imprecision of ordinary experience and judgment, it seems hardly plausible that the accumulation of experience could have yielded the principles of geometry, in the precise form in which we know and trust them. This general consideration, which Plato had urged in some form long before Kant, only gained support from nineteenth-century advances in the psychology of perception, particularly the awareness of the divergence between the space of visual perception and the space described by Euclidean geometry. The kind of empiricism that provided a plausible alternative to Kant was one that acknowledged the force of his argument, and so recognized the principles of geometry as somehow practical and ideal at the same time. This meant that the principles themselves could not be seen as inductive generalizations, but did not rule out the possibility that their empirical origins could be revealed.

Kant's theory was the starting point for Helmholtz's work on the empirical origins of geometry. He did not attempt to deny that geometrical principles have the peculiar dual character identified by Kant, as both empirical and formal – as one might say, both synthetic and a priori. Instead, he took this as an empirical fact requiring an empirical explanation. Understanding the possibility of non-Euclidean geometry, in particular, required a clear understanding of why Euclidean geometry seemed to provide such a compelling picture of actual space. By the middle of the nineteenth century, it was obvious that non-Euclidean geometries were mathematically possible, their propositions provable from their axioms by the same constructive means as in Euclidean geometry. Therefore Kant's own position was strictly impossible: Euclidean geometry was not the unique framework in which the objects of geometry could be constructed, or geometrical propositions rigorously proven. It did seem possible to retreat, however, to the position that Euclidean geometry was distinguished from other geometries that were merely mathematically possible by the fact that it could be "visualized" or "imagined." This raised again, in a different form, the question about science and metaphysics that had arisen between Kant and Leibniz: on what grounds can philosophy address the nature of space on its own, independently of mathematics and physics? Or, what authority can be claimed by the mathematics or the physics of space in order to criticize a philosophical account of spatial intuition?

An answer to these questions was, arguably, Helmholtz's most important philosophical contribution. His insight was that in the context of philosophical discussion, the relevant concepts had never been carefully defined. In order to judge whether non-Euclidean geometry is really impossible to "intuit" or to "visualize," we need to understand in what sense *any* geometry may be said to be visualizable. Therefore Helmholtz gave a general definition:

By the much abused expression "to represent to oneself [sich vorstellen]", or "to be able to imagine [sich denken] how something takes place", I understand – and I don't see how one could understand anything else thereby, without giving up all the sense of the expression – that one could depict the series of sense-impressions that one would have if such a thing took place in a particular case. (Helmholtz, 1870, p. 8)

The philosophical significance of this definition has often been remarked upon.[11] It is easy to understand why it would later appear to the logical positivists as a model of epistemological analysis, for it seems to reduce the meaning of a theoretical concept to its true "empirical content." While this gloss has a grain of truth, it overlooks the character of the definition as a conceptual analysis. Helmholtz did not purport to translate the concept of "imagination" into some sort of observation language. Nor did he merely propose to *stipulate* an interpretation of the term "to imagine" that would enable him to discuss the possibility of non-Euclidean geometry. Rather, he tried to capture the way in which the concepts are actually used, whenever they really are used in an empirically meaningful way. In order to understand whether it is possible to imagine that space is curved, Helmholtz sought to make precise just what is meant by the claim that we can imagine a flat space, or any spatial structure at all. Helmholtz's analysis therefore has something of the dialectical function that we have noted in earlier cases, intended to overcome the prevailing opinion by exhibiting and critically analyzing the implicit assumptions on which it is based.

The philosophical character of this argument suggests two further points. One is that Kant himself, by binding the empirical content of geometry to the possibility of elementary constructions, prevented any recourse from this argument in the direction of some supra-empirical, transcendent basis for Euclidean geometry. If the elementary constructions are open to a further analysis along the lines proposed by Helmholtz, then the content of geometry itself will have been completely explained. To claim that there was some other kind of foundation for geometry, merely in order to resist Helmholtz's constructive argument for non-Euclidean geometry, would not be a defense of a Kantian position at all. For then the central Kantian argument against Newtonians and Leibnizians would be negated; the doubt would be created again, whether our procedures for proving geometrical propositions are revealing the true nature of things, and so whether the propositions are truly necessary and universal.

The second point is that, as an argument against the Kantian position on its own terms, Helmholtz's is something entirely distinct from the sort of psychological argument that he directed against the nativist theory of perception.[12] On the nativist theory, awareness of the structure of space arose from the immediately spatial character of our visual sensations; Helmholtz's critique was an empirical study of the relations between visual stimuli and perceptual judgments, to reveal the degree to which spatial awareness must be gradually acquired, and spatial judgments must depend on "unconscious" inductive inferences from experience. In short, the aim of that critique was simply to show that spatial knowledge is not innate. But such a critique would have little force against the Kantian view, which was equally opposed to nativism: space for Kant was not given in the content of sensation, as the nativist view suggested, but, again, belonged to the form of sensibility. The concept of space is

acquired, not, indeed, by abstraction from the sensing of objects (for sensation gives the matter and not the form of human cognition) but from the very action of the mind, which coordinates what is sensed by it . . . Nor is there anything innate here except the law of the mind, according to which it joins together in a fixed manner the sense-impressions made by the presence of an object (Kant, 1770, p. 400)

In other words, to give an empirical explanation of Kant's view is not to explain any innate capacity to sense spatial localization. It is to explain how the "successive synthesis of the productive imagination in the generation of figures" comes to follow certain laws, and how it is that the application of those laws comes to conform to Euclidean geometry. So, at least, Helmholtz interpreted Kant's view of geometry as concerning the form of outer intuition:

By that he appears to mean, not merely that this form that is given a priori has the character of a purely formal scheme, in itself devoid of any content, and into which any arbitrary empirical content would fit. Rather, he appears also to include in the schema certain details, whose effect is precisely that only content that is restricted in a certain lawlike way can enter it, and become intuitable for us. (Helmholtz, 1870, p. 4)

Thus, while it was undoubtedly influenced by his empirical–psychological study of spatial perception, then, Helmholtz's account of "imagination" is not such a study. Rather, it is a philosophical analysis of the assumptions upon which the Kantian "productive imagination" implicitly relies. Instead of trying to reduce our geometrical intuitions to inductive generalizations, Helmholtz's analysis reveals their dependence on deeper assumptions which, themselves, arise from empirical conditions – "the facts which lie at the foundations of geometry."

The results of Helmholtz's analysis are now relatively familiar topics of philosophical discussion.[13] The empirical sources of our knowledge of space, he found, are the propagation of light rays and the motions of rigid bodies; these provide us with the notions of straight line and congruence that underlie both our intuitive sense of direction, distance, and size, and our geometrical sense of the possibility of Euclidean constructions. In the case of light rays, the connection seems straightforward enough. Our ability to "imagine" the production of straight lines in arbitrary directions for arbitrary distances, and so to represent to ourselves the Euclidean propositions involving straight lines, arises from our experience of light rays, or, more precisely, of treating lines of sight as straight lines. Experience of this kind begins practically at birth, as we learn by trial and error to reach for objects in our visual field, or to shield our eyes from light. If it seems evident, for example, that two straight lines cannot enclose a space, or that a given pair of lines must intersect if extended sufficiently, this is because we have well-established expectations about the relations between lines of sight and motions of observable bodies. Our "productive imagination" behaves according to rules because it exploits our familiarity with a stable regularity of nature. On this basis, Helmholtz could defend the intuitive plausibility of non-Euclidean geometry: if our sense that the world is Euclidean depends on the behavior of light rays as the physical counterparts of straight lines, then we can picture a non-Euclidean world by picturing light rays that behave like the straight lines of a non-Euclidean space. For example, we can predict the visual sensations that we would have if two lines could indeed enclose a space, since in certain cases we could see "through" an obstacle to some object lying behind it.[14] We can even produce such sensations, using lenses that refract light in such a way that they strike the eye as if they had followed the straightest lines of a curved space. Our Euclidean expectations would cause us to misjudge the distances of objects in our visual field, but we would quickly develop a new pattern of expectations and judge correctly (Helmholtz, 1870, p. 27). In short, as long as light travels on the straightest lines of space, it is a straightforward matter to arrive at a visual picture of a non-Euclidean world. And this means that our knowledge of geometry, Euclidean or non-Euclidean, involves an adjustment to conditions in the physical world.

The case of rigid bodies involves a subtler and perhaps deeper connection between geometry and experience. With only the image of an external world, provided by the incidence of light on the retina, we could never grasp the extension of space in three dimensions as we do, much less its geometrical properties; moreover, blind persons do develop a conception of the extent and the structure of space. From the first point it follows that light rays are not sufficient for our grasp of the extension of space, and from the second it follows that they are not necessary. What is most urgently required is the notion of displacement, arising from the possibility of freely moving the body — and therefore our point of view — between arbitrary positions in space. Like retinal images, the sensations of motion (muscular innervation) were objects of Helmholtz's special psychological study, both for movements of the body and for movements of the eyes to bring retinal images into focus. But the true foundation of Helmholtz's account was, here again, not an empirical investigation but a conceptual analysis, in a straightforward sense recently captured by Demopoulos: "the practice of recovering a central feature of a concept in use by revealing the assumptions on which our use of the concept depends" (2000, p. 220). The notion to be analyzed was the notion of space itself: what is it in our experience that we identify as the experience of space? More precisely, among the innumerable changes that we observe in our sensory environment, how do we come to distinguish certain changes as changes of spatial relation? Kant, having identified the necessity of spatial relations as the condition of the possibility of other kinds of relation, was content to take the former for granted. But Helmholtz saw the possibility of a further analysis. His question turned out to have a strikingly simple answer: spatial changes are just those that we can bring about by our own willful action; the movements that effect spatial changes can be done, undone, and combined arbitrarily (Helmholtz, 1878, pp. 225–7). In fact these are just the features of spatial displacements that allow us to treat them as forming a group, the "group of rigid motions." Poincaré's group-theoretic account of space

(Poincaré, 1902, pp. 76–91) is only a psychologically more detailed, and mathematically more precise, articulation of Helmholtz's brief analysis. The same analysis, moreover, formed the basis of Klein's "Erlangen" program for the general classification of geometries by the groups of transformations that preserve their fundamental invariants (see Klein, 1872). In this way spatial geometry came to be seen as a special case of the general theory of structures and their automorphisms (structure-preserving maps).[15]

From an epistemological perspective, there is another striking implication of Helmholtz's analysis: the group structure of spatial displacements is a crucial part of what permits us to distinguish space from time. The possibility of changing position and, to all appearances, returning again to the former position – restoring the sensible world to its former state, as it would seem – fosters our sense that space is completely independent of time. Even when we've accepted the motion of the Earth, and therefore the impossibility of returning to the same absolute position in space, the mere possibility of cancelling out changes in relative position indicates the independence of space. Of course such operations take time, and eventually it would become evident that the role played by time could not be set aside after all. But the group structure, implicit in the apparent existence of an inverse for every spatial displacement, allows us to think that this elapse of time is completely incidental.

Together, then, the concepts of the rigid body and the line of sight (the light ray) provide the geometrical notions of congruence and straightness with whatever intuitive content they actually have. By the same token, they provide the principles of geometry with a basis in empirical fact. Helmholtz was able to formulate this idea as a mathematical proposition, and so show that the assumption of free mobility is sufficient to derive the existence of a Pythagorean metric for space (Helmholtz, 1868). Riemann (1867) had already shown that the Euclidean metric implies free mobility, but Helmholtz's proof, in the opposite direction, was intended to be more epistemologically suggestive; it expresses more directly the idea that free mobility is the physical source of our geometrical knowledge.[16] It has been rightly noted that Helmholtz speaks simplistically of "the facts that lie at the foundations of geometry" (see Helmholtz, 1868), whereas Riemann refers more circumspectly to "the hypotheses." In this way Riemann indicated his awareness that the rigid body and the light ray are to a certain extent idealizations, accurate enough in familiar circumstances but likely to break down in application to "the immeasurably large" or "the immeasurably small." There can be no doubt that, in comparison with Riemann's profound and general conception of space – the basis of modern differential geometry – Helmholtz's view appears extremely narrow (see Stein, 1977). In Helmholtz's defense it may be said that he was not merely taking for granted physical concepts that he might have questioned. He was also trying to emphasize, in opposition to Kant, that the conditions of the possibility of geometry include not only the constraints upon our own sensibility – the form of outer intuition – but also the material constraints imposed by the world. That is a world in which there are bodies that we may treat as approximately rigid, and in which the propagation of light is lawful enough for accurate judgments of direction and distance. In other words, the empirical nature of geometry appears not only in the possibility of non-Euclidean geometry, as a possible outcome of empirical measurements of curvature, but in our ability to conceive of a world in which geometry as we know it would not be possible at all.

3.6 THE CONVENTIONALIST CRITIQUE OF HELMHOLTZ'S EMPIRICISM

Helmholtz provided a convincing argument that geometry has a foundation in empirical fact: our conceptions of space originate from the physical possibility of free mobility, and our ability to imagine and to perform geometrical constructions is contingent on certain regularities that hold to some good approximation. Yet Helmholtz's empiricist view, in retrospect, appeared to have overlooked some peculiar features of geometrical principles. Supposing that the principles of geometry are in some way tied to the empirical facts, can we conclude that they are therefore, as Helmholtz thought, empirical principles? The question is not whether they are, instead, hypotheses, as suggested by Riemann; their peculiarity does not lie in the fact that they are rough approximations or subject to eventual revision. On the contrary, it lies in the fact that, despite their seemingly factual content, there is something about them that appears to place them beyond empirical control. Helmholtz seemed to think it a matter of fact that there are bodies that may be moved about without change of shape or dimension. But if a body failed to satisfy this condition, how would we determine this empirically? Only by comparing it with another body that we suppose has remained rigid. Similarly, how would we know empirically that light failed to travel in a straight line? Only by comparing it with some physical object or process that we suppose is truly straight. In either case, we are at the beginning of an infinite regress, at every stage comparing one physical process to another that is supposed to have the desired property. The natural conclusion was drawn explicitly by Poincaré: the principles that Helmholtz alleges to be facts are in reality "definitions in disguise." That light travels in a straight line is a definition of the straight line, not an empirical claim about light. That certain bodies remain congruent to themselves under certain motions is not an empirical fact about those bodies, but a definition of congruence. Such principles, therefore, cannot be arrived at by inductive arguments. As Poincaré pointed out, their subject matter is the way in which certain concepts are to be applied. Therefore they can only be fixed by stipulation.

Poincaré's conventionalism is the subject of a large literature,[17] to which this discussion can only add a few important points. One is that, in spite of himself, as it were, Helmholtz already took note of this curious aspect of the principles of geometry. For one thing, he raised the possibility that a neo-Kantian might defend Kantianism against his arguments by regarding the concept of rigid body as a "transcendental concept, formed independently of actual experience." In that case, the principles of geometry would be immune to empirical refutation, because "one would have to decide according to them alone whether any given natural body is to be regarded as rigid" (Helmholtz, 1870, p. 30). In other words, Helmholtz's argument would be evaded, non-Euclidean geometry would be empirically impossible, if we were certain that only those bodies were rigid whose behavior conformed to the principles of Euclidean geometry. In fact a version of this idea was put forward by Hugo Dingler in the early twentieth century, as part of a general criticism of Helmholtz's view. According to Dingler, Euclidean geometry is presupposed in the construction of our measuring instruments and so has a privileged status, unassailable by any results that we might obtain by manipulating those instruments (see Dingler, 1934, also Carrier, 1994). As Helmholtz noted, however, such a response would involve a serious departure from Kant. For then the principles of geometry would no longer be synthetic a priori, but only analytic. That would defeat the central point of Kant's account of geometry. To Helmholtz, the resolution was fairly straightforward: the principles of geometry are synthetic, but their empirical content comes from the principles of mechanics and optics.

We can see in other remarks by Helmholtz, however, some recognition that this simple answer masks some complications. The principles of mechanics and optics cannot settle the question of geometry in the way that Helmholtz suggested, if they themselves invoke the principles of geometry. If we know that light travels in a straight line, or that our measuring instruments move without changing their dimensions, then we can determine the geometry of space by experiment, but, as we have already seen, we could not be said to know such principles by experience. In fact what we do know from experience, on these points, Helmholtz knew better than anyone: not only that there are motions that we can distinguish as spatial displacements, but also that a line of sight between two places corresponds to the direction of shortest motion between them. That a line of sight represents the straight line of geometry is an obvious interpretation, but it is still a kind of interpretation. Helmholtz even seems to suggest as much, when he raises the question of distinguishing facts from definitions in the foundations of geometry:

In my opinion this question is not so easy to answer, for in geometry we deal constantly with ideal structures, whose corporeal representation in the actual world is always only an approximation to the requirements of the concept, and we first decide whether a body is rigid, its sides flat and its edges straight, by means of those same propositions whose factual correctness the examination is supposed to prove. (Helmholtz, 1868, p. 618)

This applies most conspicuously to the case of the light ray and the rigid body, but more pointedly to that of the rigid body because of its inseparability from the notion of congruence. Hence any attempt to define the empirical meaning of the principle of rigid motion has to appeal to congruence, and vice versa: "we have no criterion for the rigidity of bodies and spatial forms except that, when juxtaposed to one another at any time, in any place, and after any rotation, they always exhibit the same congruences as before" (Helmholtz, 1870, p. 29). Regarding the Euclidean postulates, Helmholtz clearly recognized that to use them as *criteria* for rigidity and straightness, for example, would amount to treating them as analytic instead of synthetic. But the principle of free mobility evidently poses the same problem, and Helmholtz left it for others (especially Poincaré) to make the problem explicit. In short, as we saw in the case of absolute time, the principle of free mobility plays the role of a definition. Its empirical content is therefore a kind of general expectation, namely, that for various methods of measurement, the better they approximate a certain ideal of rigidity the better they will agree with one another.

Indeed, one might gather from Helmholtz's remarks that this definitional character is pervasive in the principles of mathematics. His paper on the foundations of arithmetic sets out to treat arithmetic as he had treated geometry, that is, as a formal science developed from empirical principles. In other words, it would appear, arithmetic is to be traced to its empirical foundations just as geometry had been; we might expect him to identify the physical facts that underlie our practices of counting and measurement.

But if this was difficult to accomplish in the case of geometry – if, as we saw, the apparently physical principle of free mobility was hard to represent as a straightforward empirical principle – it was still more difficult for the principles of arithmetic. Helmholtz acknowledged this explicitly in his analysis of arithmetical principles. Concerning the axioms of arithmetic, Helmholtz writes, "The first axiom – 'If two magnitudes are both alike with a third, they are equal to each other' – is therefore not a law with objective significance; it only determines which physical relations we may recognize as equality" (1887, p. 380). Then, in defining "physical connections" that represent the addition of physical magnitudes, Helmholtz states even more explicitly that he is not dealing with ordinary empirical laws: "we should not wonder if the axioms of addition are verified in the course of nature, since we recognize as addition only those physical connections which satisfy the axioms of addition" (1887, p. 384). If this is true, then arithmetic, like geometry, is not derived from facts of experience. Rather, it consists in the application of certain concepts to experience, or, more precisely, the search in experience for instances of those concepts. This by itself is by no means a novel idea on Helmholtz's part. The notion that mathematical concepts, instead of being derivable from experience, are *criteria* by which phenomenal objects are judged, or ideals to which they are compared – goes back at least to Plato. What is novel is the implication that the concepts are first defined by the axioms in which they appear, so that they do in fact function as disguised definitions. Only Helmholtz does not appear to have appreciated the significance of this fact for his original empiricist aims. Consequently he appears not to have noticed that the principles of mathematics emerge now as analytic principles, rather than as the synthetic a-posteriori principles that he had thought to exhibit. In the case of geometry, the very possibility of such an interpretation was discounted because the laws of mechanics, he assumed, provided the factual grounds for a choice among possible geometries.

This last point is the lesson Helmholtz draws from his celebrated examples of non-Euclidean spaces. If we observe the reflections of ourselves, and of the movements of our measuring instruments, in a spherical mirror – then as now a popular garden ornament – we would find that what we think of as rigid motions produce systematic distortions of size and shape, and our straight lines would be reflected as curves (Helmholtz, 1870, pp. 24–5). Yet objects that we found to be congruent would also be congruent on the sphere; since all bodies would be equally distorted, the coincidences of our measuring rods with the objects that we measure would not be disturbed. If we could imagine inhabitants of this spherical world, attempting geometrical measurements, we must admit that the distortions that we observe would be completely imperceptible to them, and that – because all geometrical coincidences would be preserved – the bodies that look distorted to us must look rigid to them. Moreover, our world would necessarily appear to them a distorted reflection of theirs. "And . . . if the men of the two worlds could converse together, then neither would be able to convince the other that he had the true, and the other the distorted situation" (Helmholtz, 1870, p. 25).

From the perspective of the twentieth century, Helmholtz's argument has been read as showing the "relativity of geometry": it shows that there can be completely incompatible alternative descriptions of the same situation, and no principled way of choosing among them (see, for example, Van Fraassen, 1989). Helmholtz's thought experiment is also an obvious ancestor of Einstein's view, that our empirical knowledge of geometry consists entirely of "verifications of . . . meetings of the material points of our measuring instruments with other material points" (Einstein, 1916, p. 14). Since whatever pairs of objects are congruent in a picture must be congruent in its distorted image, the "point-coincidences" determined by such congruences are the only facts agreed upon from both perspectives, the only facts that are verifiable independently of one's choice of perspective. For Helmholtz, however, the lesson does not concern relativity at all. Rather, it concerns the inseparability of geometry from mechanics. We could not dispute the views of the sphere inhabitants without appealing to "mechanical considerations" – that is, we can argue with them only on the assumption that we share a common world of mechanical laws, so that we can determine which picture expresses the true spatial dispositions of bodies and light rays, and which picture is only a distorted image. The issue of relativity arises between completely different worlds, not between different descriptions of the same world; since the latter must by hypothesis have a single set of physical laws, its geometry must be a matter of empirical fact. Physicists who live in the same world, unlike those regarding each other from outside as in the example, will inevitably find a common way to determine its physical laws.

As Poincaré's considerations reveal, however, such a conclusion involves some question-begging. The sphere dwellers might agree to resolve our differences on mechanical grounds, yet reject our interpretation of the fundamental relations between geometry and mechanics, and thereby reject our understanding of basic geometrical concepts. If they insist that light in their world is traveling on straight lines, or that their measuring instruments are the truly rigid bodies, how then shall we convince them otherwise?

In short, their conception of geometry, like ours, must depend on what physical objects they take to be rigid, and what paths they take to be straight. This is no mere artifact of Helmholtz's example, with two disconnected worlds; it is not impossible to imagine a physics developed in our own world, but from a different starting point and along independent lines, with (for whatever reason) different conceptions of which bodies are appropriate measuring instruments; the adherents of that view would have trouble convincing us of any empirical facts about space. It follows that neither their position nor ours could have any firmer ground than the neo-Kantian view that Helmholtz had proposed. Either the principles of geometry, or the definition of the fundamental geometrical quantities, must be set down a priori. If the former, then which bodies are truly rigid or which paths straight becomes a matter for empirical investigation – for the truly rigid bodies and the truly straight lines will be those that accord with the chosen geometry. If the latter, then the comportment of bodies and straight lines will reveal the geometry of themselves.

This brings us to the second important point about Poincaré's conventionalism. Whether he was right to say that these a-priori principles are only conventions, and even what he really meant by saying so, are perhaps open to question. But his identification of these principles, and of their roles in our conceptions of space and time, must be accounted a genuine philosophical discovery. One may speak of a-priori principles, in physics at least, as if they might originate as empirical principles, and have the status of a priori conferred upon them by our act – as if an empirical generalization might be, as Poincaré put it, "exalted" to become a postulate.[18] But the essence of Poincaré's view is not that we can agree to treat empirical principles as "absolute" and unrevisable, or to treat approximately verified principles as exactly true. It is, rather, that some of the most important principles that we take to be empirical simply are not – that they are, by their very form and content, "definitions in disguise." The concepts about which they seem to inform us are in fact defined for us by those very principles. The principles therefore do not make assertions about actual states of affairs. Rather, they establish how particular concepts are to be empirically interpreted. This is not merely a fact about geometry, but is true of any science that applies mathematical structures to experience, including theoretical physics in general. That force is proportional to acceleration cannot be, in itself, true or false, unless we assume that there is something prior to Newton's physics that gives us an independent definition of force; the function of this principle in the Newtonian framework is to fix the meaning of the concept of force, by associating it with a measureable feature of a body's motion. We could hardly say that force seems to be nearly proportional to acceleration, and that we have adopted the convention that it is exactly so; we know that it is exactly so because Newton's law imposes this definition as the measure of force. Any imprecision, in the measure of a particular force from a given acceleration, only requires us – by definition – to seek out forces contributed by yet-unnoticed bodies. Poincaré's analysis, then, in the best philosophical tradition, uncovers a subtle philosophical error: mistaking a definition for an empirical principle, through a failure to see the assumptions on which the content and the application of the principle really depend.

Poincaré made this most clear in his exchange with Bertrand Russell over the foundations of geometry (Russell, 1897, 1899; Poincaré, 1899a, b).[19] On Russell's account, propositions are meaningful just to the extent that their constituents are understood, and the meaningfulness of geometrical principles depends on our independent grasp of the primitive terms that occur in them. As Poincaré showed, however, a proposition like "bodies can be moved in space without change of shape" is not telling us something new about a previously understood conception of shape; rather, this proposition is partly constitutive of any understanding of "shape" that we have. The real purport of the statement is that "in order for measurement to be possible, it is necessary that figures be susceptible of certain movements, and that there be a certain thing that will not be altered by those movements and that we will call 'shape'" (Poincaré, 1899a, p. 259). For Russell, shape is just the sort of "indefinable" basic term of which we have an unanalyzable, immediate grasp, and asking for a definition of it is like asking for "the spelling of the letter A" (Russell, 1899, p. 701). For Poincaré, that allegedly immediate grasp may be relevant to the psychology of the individual. But it cannot help us to understand how a concept functions within a coherent system of principles, so that two persons who may not associate the same immediate intuition with a concept can nonetheless reason with it in accord with one another. That requires that both apply the same systematic criteria to recognize instances of the concept, and those are given only by the principle that constitutes its implicit definition. This has two notable consequences: that the vague intuitive notion can be made precise by articulation of the principle that is implicitly assumed in our use of it; and that such a principle, once isolated, allows us to treat the subject as a formal one that is independent of intuition altogether.

In retrospect, this view accords with Kant's much more than Russell's does, despite the latter's emphasis on intuitive self-evidence. Kant's shares with Poincaré's an essential anti-psychologistic emphasis; in both cases, it is formal principles that render intuition a source of systematic knowledge. Arguably, the "form of intuition" shares that emphasis not only with Poincaré's view, but also with Plato's explanation of geometry through the theory of forms. All of these acknowledge that what we are immediately given in intuition – misleading appearances, badly drawn diagrams, inaccurate estimates – is inadequate to the demands of geometry for a precise conception of space. Since we evidently do have such a conception, its foundation must be something beyond what is given, something "called to mind" by the given but with a conceptual component, an element of universality and necessity, that transcends what is given. Plato supposed that this must come from intuitions of another sort, given to a purely intellectual faculty. But Kant saw the difficulties of this supposition, and saw that the formal principles, rather than being themselves objects of intuitive knowledge, must somehow order the combination and the "successive synthesis" of sensible intuitions. So these transcendental principles, conditions under which intuitions can provide objective knowledge of relations in space, cannot transcend the intuitions whose order they constitute. It was Poincaré who understood that if there is such an order, or "form of intuition," it must be implicit in the concepts that we impose upon intuition. Those concepts are implicit in the rules that guide our intuitive practice.

It is true, as we have seen, that Helmholtz glimpsed this possibility, if somewhat dimly. But he never saw quite clearly enough its implications for the empiricist view that he was advancing. With Poincaré, we see the explicit realization that principles such as the principle of free mobility, and other fundamental constitutive principles, belong to a distinctive type. Though stated in the grammatical form of synthetic propositions, they are really interpretive principles that assign meaning to particular concepts. And this is why they have that aura of necessity that Kant emphasized, and even Helmholtz had acknowledged. But, as Coffa expressed Poincaré's view, "convention, semantically interpreted, is merely the opposite side of necessity. In the range of meanings, what appears conventional from the outside is what appears necessary from the inside" (Coffa, 1991, p. 139). What protects such principles from revision is not their apodeictic certainty, but our decision to treat them as unrevisable.

3.7 THE LIMITS OF POINCARÉ'S CONVENTIONALISM

Poincaré's account of the foundations of geometry seems to imply a thoroughgoing relativity of space and of spatial knowledge. Where Helmholtz had stubbornly held on to some form of empiricism, for example in his account of the spherical mirror, Poincaré saw that such an implication could not be avoided. It was he who pointed out that as far as empirical facts are concerned, there is no distinction at all between the world as we think we know it and one with completely distorted metrical relations – provided only that everything is equally distorted, and that all coincidences are preserved between material points. Thus Poincaré explicitly anticipated what the logical positivists would call "the relativity of geometry" (see Schlick, 1917, chapter 3). What we learn directly from intuition is only which bodies appear to coincide, and any further knowledge of geometrical relations is possible only if certain conventions are imposed. In this way Poincaré took the criticism of Kant a large step beyond Helmholtz. There, the point had been that the intuitive–constructive basis of geometry is actually weaker than Kant had required, that is, too weak to single out Euclidean geometry among the geometries of constant curvature. Poincaré pointed out that the underlying principle even of this general class of geometries, the principle of free mobility, has a quantitative content that goes beyond simple intuition; a truly intuitive geometry must be still more general, eliminating even the simple requirement of free mobility, and including only what is accessible to direct intuitive verification, without physical (metrical or quantitative) assumptions. This is why Poincaré thought that the only truly intuitive geometry was *analysis situs*, in which only topological distinctions matter and measurement is out of the question (Poincaré, 1913).

But the difference between Poincaré's view and its twentieth-century successor is significant and illuminating. In the context of general relativity, as understood by Einstein and Schlick, the relativity of geometry has an immediate significance for metrical geometry: the underlying "amorphous" space is represented by an arbitrary Riemannian manifold, assumed to have no more intrinsic structure than its differentiable structure (see Chapter 4, later). Physical structure, more precisely the metrical structure that is to play the role of the gravitational field, is imposed by two stipulations: first, that special relativity holds in the infinitely small, i.e. at any point the metric is Minkowskian; second, that over finite regions, the metric depends on the mass distribution in accord with Einstein's equation. From Poincaré's insight into the epistemology of geometry, it seemed, one could infer that space is in itself an empty notion that must get its physical content by our decision.

In Poincaré's view, however, such a space is essentially pre-geometrical or, perhaps more precisely, non-geometrical. The only geometry it can be assumed to have is "intuitive" geometry in Poincaré's peculiar sense, which is not really geometry at all. What Poincaré thought of as genuine geometry concerns a much more restricted kind of space, the space whose properties we come to know by way of the group of congruence transformations: "Geometry is merely the knowledge of the mutual relations of these transformations, or to use mathematical language, the study of the group formed by these transformations, that is, the group of motions of solid bodies" (Poincaré, 1902, p. 99). This definition evidently admits only a special case of Riemannian geometry in general, that is, again, only the geometries of constant curvature. The reason for this was that to Poincaré, all other possible geometries were analytic rather than synthetic: because the principle of free mobility does not apply, they are not knowable by classical geometrical constructions. They are objects of mathematical study as formal systems, but not susceptible of empirical interpretation as structures for space. Intuition, then, remained as central to space for Poincaré as it had been for Kant, at least to the extent that, for both of them, the intuitive constructive procedures were conditions of the possibility of spatial knowledge.[20]

This last point sheds some further light on the nature and limits of Poincaré's conventionalism. At least as far as the physical content of geometry is concerned, he was not really a conventionalist at all; unlike the positivists who drew encouragement from his remarks, he did not view spatial geometry as inherently empty formalism. The content of geometry was not imposed by convention, but revealed by the conceptual analysis that identified it with our experience of free mobility. Poincaré's conventionalism arises, in fact, precisely from the way in which the content of geometry is fixed. The principle of free mobility, again, only establishes that space has one of the geometries of constant curvature. To Helmholtz, it seemed obvious that we could appeal to physics for an empirical decision among these; to Poincaré, such an appeal is *inherently* conventional, because it involves principles that are inherently *extrinsic* to the concept of space. Any principle we might adopt, such as that light travels in a straight line, must be a dynamical principle, i.e. a principle involving *time* as well as space. Helmholtz or Riemann might object, at this point, that the laws of physics have some claim to authority in these matters, since geometry is in the end dependent on mechanical assumptions. For Poincaré, the dependence was in the other direction: physics could not even begin – its basic dynamical principles could not even be formulated – unless a spatial geometry was already given (see Friedman, 1999b).

This appears to be an obvious fact about Newtonian mechanics, which had always taken for granted a background Euclidean space against which basic quantities such as position and velocity could be defined. In Poincaré's view, it was only an instance of a more general fact about the mathematical sciences, namely that they stand in a kind of hierarchy of dependence (see Friedman, 1999b). Physics could not begin to define its concepts except against an assumed background of geometry; geometry required an assumed background of arithmetic; arithmetic in turn required the background of a general theory of magnitude, and so on. There could be no question of a principle of physics determining the nature of spatial geometry, any more than a principle of geometry could determine the nature of arithmetic; in either case it is impossible, because the former principle implicitly takes the latter for granted. At the level of geometry, it happens that the defining principle (free mobility) is insufficient to determine the structure completely, beyond the fact that it must be of constant curvature. Mathematicians had established that Newtonian physics could be compatible with non-Euclidean spaces of constant curvature, but, as Poincaré recognized, this only confirmed the impossibility of appealing to physics for a decision. In order to proceed from geometry to physics, then, some convention must be adopted that must borrow from physics itself. That this curious fact might turn out to undermine the ordering of Poincaré's hierarchy does not seem to have occurred to him – even after it happened, when the electrodynamics of moving bodies led to a new understanding of spatial geometry (see Chapter 4, later).

Poincaré's view makes an instructive comparison with Riemann's (1867). Where the conceptual analysis of Poincaré and Helmholtz made free mobility the fundamental defining principle of spatial geometry, Riemann's analysis identified a much more general conception of space as a "multiply-extended manifold," a notion that might apply to any collection of elements that could be specified as "locations" with respect to some number of continuously varying magnitudes. He identified the metric of a space as a function of these locations – the differentials of the coordinates in each of the several dimensions – and some arbitrary and variable coefficients, in an expression for the infinitesimal distance ds^2 that is now fairly familiar:

$$ds^2 = \sum_{\mu,\nu} g_{\mu\nu} dx^\mu dx^\nu$$

This is not the place for an extended survey of the subject,[21] but we can see that in the three-dimensional case, the differentials form the matrix

$$\begin{pmatrix} dx^1 dx^1 & dx^1 dx^2 & dx^1 dx^3 \\ dx^2 dx^1 & dx^2 dx^2 & dx^2 dx^3 \\ dx^3 dx^1 & dx^3 dx^2 & dx^3 dx^3 \end{pmatrix}$$

and the coefficients form the matrix

$$\begin{pmatrix} g_{11} & g_{12} & g_{13} \\ g_{21} & g_{22} & g_{23} \\ g_{31} & g_{32} & g_{33} \end{pmatrix}$$

Then it is easy enough to see that the infinitesimal Euclidean (Pythagorean) metric

$$ds^2 = dx^2 + dy^2 + dz^2$$

is just the special case of a Riemannian metric in which the matrix $g_{\mu\nu}$ has the form

$$\begin{pmatrix} 1 & 0 & 0 \\ 0 & 1 & 0 \\ 0 & 0 & 1 \end{pmatrix}$$

that is, where $g_{\mu\nu} = 1$ where $\mu = \nu$, and $g_{\mu\nu} = 0$ where $\mu \neq \nu$. It is also easy to see how different values for the coefficients $g_{\mu\nu}$ determine an infinite variety of metrical structures, homogeneous and inhomogeneous, including, most notably, the space-time structures of general relativity in which $g_{\mu\nu}$ depends on the distribution of mass.

Again, Poincaré knew Riemann's work well, and both of them understood that the concept of physical space involved severely restricting the more general conception. But for Riemann, as for Helmholtz, the condition of free mobility was a restriction that rested on special empirical assumptions; even more than Helmholtz, Riemann emphasized that the idea of a rigid body is one that physical objects can only approximate. More important, instead of proposing to adopt the convention that there are rigid motions, Riemann pointed out that the advance of physics would lead us to more exact notions at smaller (or possibly very large) scales. In other words, the principle of free mobility was not for Riemann a condition of the possibility of physical geometry, but an assumption that we rely upon provisionally, until we gain some deeper insight into the nature of bodies and their microscopic interactions. Then the concepts of rigid body and light ray may well be inapplicable, and the idea that space must be homogeneous – the entire picture of space as founded in a group of isometries – may prove to be a drastic over-simplification. Riemann's philosophical position, then, was a more sophisticated and forward-looking version of Helmholtz's empiricism, on which physics was acknowledged to be the source of our geometrical knowledge, and therefore authoritative over it; only Riemann imagined that physics itself might call into question the concepts it had furnished for geometry, and eventually place geometrical reasoning on an entirely new foundation. Evidently it was this philosophical attitude, as well as the mathematical formalism that he introduced, that made Riemann seem like an anticipator of twentieth-century physics.

The empiricist view, then, was that dynamical principles – principles involving time as well as space – could force revision of the spatial geometry that had been originally assumed in their development. We might say that this view acknowledges the possibility, at least, that space-time is more fundamental than space. On Poincaré's hierarchical view, in contrast, such a revision would not make sense. It would appear that special and general relativity confirmed the empiricists, since radical changes in geometry were motivated by considerations of electrodynamics and gravitation, respectively. In one sense, however, this is a slight exaggeration. In fact Einstein did not force the revision of spatial geometry directly, through any breaking down of the concepts of rigid body or light ray; rather, he inferred from the "gross" behavior of light and bodies that there were difficulties in our conception of space and time together – in the conception of simultaneity, or the separation of space and time, and in the conception of inertia, or the notion of a privileged trajectory in space and time. Inhomogeneous spatial geometry entered into general relativity only as the projection upon space of an inhomogeneous *spatio-temporal* geometry; for example, spatial curvature near the Sun, as revealed by the bending of starlight, emerged from the theory that light travels on the geodesics of a curved space-time. But Riemann's empiricism acknowledged, at least, that spatial geometry could not be isolated from the future development of physics. For Poincaré, this isolation was an important part of what defined the role of geometry in physics.

The privileged status accorded to space, and the difficulties associated with it, define Poincaré's approach to the problems of absolute space and absolute motion (Poincaré, 1902, pp. 135–42). He certainly defends the "relativity of space" against the notion of absolute space. But his reasons for rejecting it have to do, not with the dynamical equivalence of states of motion – which involves space and time together – but with the properties of space alone. The relativity of space follows from the homogeneity of space that is embodied in the group of rigid motions, since the symmetries of Euclidean space (or any space of constant curvature) make it impossible to single out privileged positions in space. As we noted in Chapter 2, however, Newtonian absolute space fully respects the *spatial* symmetries of Euclidean geometry; what it does not respect are the *spatio-temporal* symmetries of Newton's own mechanics, the Galilean group of transformations, which make it impossible to single out *the same* position at different times. In other words, Poincaré has repeated the classical relationalist confusion, between the question of a privileged position in space and that of a privileged velocity in space and time. Absolute space permits the latter but not the former, while, properly understood, Newtonian mechanics permits neither. Yet in spite of his convictions about relativity, Poincaré acknowledges that the phenomena of centrifugal force do distinguish rotation from non-rotation. He therefore concludes that we need the "fiction" of an absolute space, even with distinguished positions, as something to which rotation can be referred. Physicists are obliged to live with the philosophical embarrassment of absolute space in order to make sense of dynamics.

It would be pointless to criticize Poincaré for maintaining the need for absolute space, even though he maintained it well after the concept of inertial frame had become fairly well known (see Section 4.1 later). It is only worth noting the difficulty in which he is placed by his conception of space and its relation to physics. Because he thinks of space as completely characterized by the group of rigid motions, and therefore inherently "relative," Poincaré is not in a position to consider absolute space as part of a spatio-temporal structure that is revealed by the laws of motion. So, he wrongly concludes that by making spatio-temporal distinctions among states of motion, the laws of motion violate the relativity of space. To introduce absolute space into physics, in order to explain such distinctions, is to adopt the "fiction" that this structure is the sort of real thing with respect to which we can speak of relative motion – much as, in electrodynamics, Poincaré was content to adopt the "fiction" of an ether as the medium in which electromagnetic waves propagate, and as the reference frame with respect to which their true velocity is defined. Superficially this view would appear to resemble Kant's, insofar as the compelling epistemological arguments against absolute space – arguments founded in the very nature of space and the means by which we come to know it – have to be set aside because of the conundrum posed by absolute rotation; neither Kant nor Poincaré saw beyond the space of intuition, to the possibility of a spatio-temporal structure that would solve the conundrum and yet respect the relativity of space. But Kant's solution is, in fact, much more abstract than Poincaré's, and so much more in keeping with Newtonian mechanics. As we saw earlier, Kant did not simply impose absolute space as a privileged relative space for the explanation of absolute motion; rather, he viewed it as the end-point of the dynamical analysis of relative motion. In short, in Kant's approach absolute space has much the same role as an inertial frame in the modern approach to Newtonian mechanics, as a frame constructed by the identification of true accelerations and forces in the interactions of a system of bodies; the modern approach merely acknowledges that we have identified an infinity of equivalent spaces whenever we have identified one. Absolute space, then, on Kant's account, rests not on an explanatory hypothesis or fiction, but on a conception of true motion as something that we can determine by a constructive process. In Poincaré's approach, however, it is a kind of reification of an abstract idea, a metaphysical embarrassment that we can countenance only by not taking it seriously as an objective feature of the physical world. The harmony between the concept of space and the theory of dynamics that Kant had sought to articulate, and that finally emerged with the concept of inertial frame, was not visible from a perspective according to which the concept of space was so completely independent of physics.

Understanding this aspect of Poincaré's thinking helps us to understand another aspect that has been much remarked upon, namely, his conviction that physics would always be founded on Euclidean geometry, and that physicists would always prefer to adjust the laws of physics rather than to adopt a non-Euclidean geometry. This seems to be, at best, a careless sociological prediction, especially since he continued to make it even after special relativity had been developed; at worst, it exhibits a degree of narrow-mindedness that seems remarkable in a mathematician who contributed so much to the understanding of non-Euclidean geometry. It is tempting to blame this puzzling conviction on his conventionalism. And there is some justice in doing so: if some convention about geometry must be adopted before physics can begin, then physicists have the right to choose the simplest geometry, which is undoubtedly Euclid's. Any complications that might arise as they try to maintain it in all circumstances must be seen as mere inconveniences, costs that must be weighed against the cost of adopting a more complex convention. Poincaré obviously disagreed with Kant that there is a unique given framework for geometry, but he agreed with the Kantian principle that, by its very nature as a general framework, Euclidean geometry was inherently capable of embracing every possible phenomenon, and that the problem of explaining away what might seem to be non-Euclidean relations was just part of the scientific task that any such framework imposes. This is no different from the manner in which Newtonian mechanics imposes the task of explaining away anything that seems to be an unbalanced force, by discovering some previously unknown mass that is interacting with whatever system we are studying. Only some form of a priorism, it would seem, would permit such a confidence, and the version of a priorism that lay behind Poincaré's confidence was conventionalism.

Yet we might also expect that conventionalism, by itself, would encourage a more casual attitude toward the changing of geometrical frameworks, a recognition that a judgment of convenience is probably provisional and likely to be revised under the pressure of new circumstances. Poincaré himself eventually spoke of special relativity and Newtonian mechanics in direct comparison, as two possible conventions regarding the group of spatio-temporal transformations, either of which can be defended on some pragmatic grounds or other. But he continued to maintain that physicists would ultimately stay with the older convention. More lies behind this conviction than just conventionalism; it is Poincaré's particular kind of conventionalism, combined with his particular view of the privileged status of space. The theory of space will not be overturned by principles of physics, because space is exhaustively defined for us as a pre-physical notion, and because, therefore, the transition from geometry to physics must always introduce extraneous elements into the concept of space. That geometry had always involved such elements, and that our changing understanding of those elements and their implications for geometry was essential to the growth of geometry as a science, was an empiricist conviction that Poincaré never took to heart.

3.8 THE NINETEENTH-CENTURY ACHIEVEMENT

The nineteenth century, then, developed a remarkable degree of insight into the foundations of empirical geometry, starting from Kant's conception of it as expressing the structure of spatial intuition, and eventually revealing its dependence upon physical assumptions about the dispositions of bodies and light rays. In the process they developed a new understanding of the a-priori aspect of geometry, showing it to be neither synthetic nor analytic in Kant's narrow sense; its principles were neither synthetic statements about real states of affairs, nor mere analyses of what was "contained in" particular geometrical concepts. Indeed, this very notion of the analytic, as Kant understood it, was inadequate to account for concepts that are implicitly defined by the axioms of a theory; for the somewhat vague question of what elements are contained in a given concept, one could now substitute the question of how the concept functions within a given framework of principles. But this question suggests that such a concept has no foundation beyond its role in an arbitrary formal structure, whereas the analyses of Helmholtz and Poincaré suggested something more than that: that the principles of physical geometry revealed the way in which certain concepts function as criteria in the organization of our experience, making our pre-systematic notions – e.g. of spatial displacement – the basis for a systematic knowledge of space as a mathematical structure. By virtue of that function they had to be considered a priori rather than empirical, but their dependence on contingent features of the world, and on our ability to recognize and exploit them, meant that they could not be imposed by arbitrary convention. In a certain sense they could be said to be discovered: not as directly revealed by empirical evidence, but as implicitly guiding a set of pre-systematic empirical practices. The association of geometry with rigid bodies arose, in other words, not by any conventional choice, but by the conceptual analysis of our empirical knowledge of space.

The impact of the nineteenth-century accomplishment was obscured, I think, by its bond with an intuitive picture of space and time, that is, with a conception of space as separate from time, and of spatial measurement as defined by intuitively obvious procedures. In effect, this work was confined by the original post-Kantian project, of accounting for the intuitive–constructive picture of geometry by an analysis of its empirical roots. Riemann's approach, it is true, stood outside of these confines, based as it was on a far more general conception of space, of which the set of intuitively constructible spaces represented an extremely special case. In doing so he went as far as possible toward abstracting a general conception of space from any possible intuitive picture of it, which must then be seen as resulting from the imposition of very special assumptions about measurement upon this general conception. And he raised the possibility that the intuitive picture, or, more precisely, the principles assumed by measurement, are merely rough approximations to the principles that really do determine the geometry of space. But this was only a partial emancipation from the intuitive picture, since it represented the latter as an approximation to a more accurate picture. A complete emancipation was required only when physical geometry had to confront dynamical principles – spatio-temporal principles – that called into question the very notion of space, as something independent of and separable from time. The task for the twentieth century was to find, for unintuitive or counterintuitive principles taken from electrodynamics and gravitation, a connection with space-time geometry that was as direct and revealing as the connection between rigid motion and spatial geometry. We will see in the next chapter how this task was eventually completed.

NOTES

1. The influence of Euler on Kant's development is discussed in detail by Friedman (1992), especially chapter 1, section II. See also Euler, "Réflexions sur l'espace et le temps" (1748, pp. 324–33); *Theoria motus corporum solidorum* (1765, 2:79–85).
2. For a helpful guide to these commentaries, see Earman (1989, chapter 7).
3. See Friedman (1992, chapters 3 and 4).
4. In explaining the system of pre-established harmony, Leibniz writes, "Thus as soon as one sees the possibility of this hypothesis of agreements [*Hypothese des accords*], one sees also that it is the most reasonable one and that it gives a marvellous idea of the harmony of the universe and of the perfection of the works of God . . . In addition to all these advantages which make this hypothesis recommendable, one can say that it is something more than a hypothesis, since it seems hardly possible to explain things in any other intelligible manner, and since a number of great difficulties which have exercised minds up to now seem to disappear of themselves when one has understood it well" (1695[1960], pp. 485–6). Even if it is "more than a hypothesis" in the sense implied, it is unquestionably a hypothesis in the sense of Newton or Kant, that is, a theory recommended by its ability to give a qualitative explanation of some known phenomena – not by any argument that can be given either "from the phenomena" in Newton's sense, or from the conditions of the possibility of experience in Kant's.
5. This section, even more than the rest of the present chapter, is heavily indebted to Friedman (1992). The reader is referred to that work for a more careful and complete discussion of Kant's interpretation of Newton.
6. See Friedman (1992). In the late nineteenth century, Maxwell (1877) recognized that this implication of Newton's laws embodied a kind of relativity beyond Galilean relativity, in which acceleration, though not rotation, had to be regarded as partly arbitrary. For this reason Earman (1989) named a structure in which rotation could be distinguished from non-rotation, but uniform acceleration is indistinguishable from rest, "Maxwellian space-time." But it is clear that Maxwell saw the situation much as Newton did, as revealing a practical difficulty in the large-scale application of the Newtonian laws rather than a deep difficulty of principle.
7. See Friedman (1992, especially chapter 3, section II) for a close analysis and an illuminating interpretation of Kant's reasoning on this point.
8. This idea was revived in the twentieth century as a resolution to the absolute–relational controversy. It seemed possible to admit the reality of absolute acceleration, but to deny that it must be referred to absolute space or space-time, by interpreting it as a kind of intensive magnitude, something whose metaphysical reality is internal to itself, so to speak, and doesn't need to be defined

by its relation to absolute space (see Sklar, 1977). The strategy, in other words, was to deny that absolute motion must be regarded as a species of relative motion, and to assert that it has an independent metaphysical standing. As a metaphysical hypothesis, the notion that there is a quantity called "absolute acceleration" that is intrinsic to bodies and does not depend on space and time is difficult to rule out. Kant's question is simply whether such a hypothesis can ever be anything but an arbitrary construction. For the conditions for defining such a quantity as a physical magnitude depend on its representation in space and time. It is not a question of the ultimate ontological ground of the quantity, but of the very meaning of the corresponding concept, or more precisely, our ability to attribute any content to it.

9. See Friedman (1990, 1992). But see also DiSalle (1990).

10. For a wider survey of empiricist and other views, see Torretti (1977, chapter 4).

11. See, for example, the notes and comments to Helmholtz (1870) by Paul Hertz and Moritz Schlick, in Helmholtz (1921); Coffa (1991, chapter 3); Friedman (1999c).

12. For further discussion of the contrast between the Kantian view of space and the nativist theories of spatial perception, see DiSalle 2005.

13. For some discussion of Helmholtz on this point, and its further influence on the philosophical literature, see (e.g.) Reichenbach (1957), Torretti (1977), or Carrier (1994).

14. Einstein's (1916) prediction of the bending of starlight, and the phenomenon of "gravitational lensing" that emerged in the later development of general relativity, are now familiar natural examples of essentially the same effect.

15. For further commentary see Torretti (1977), Friedman (2002a).

16. Riemann (1867) showed that if the metric of space is a quadratic function (a generalized form of the Pythagorean metric) of the coordinate differentials, it is invariant under arbitrary changes of position. Helmholtz (1868) showed, conversely, that if length is a function of coordinates that is invariant under arbitrary changes of position, the spatial metric has the Pythagorean form. This implies that the requirement of free mobility is equivalent to the requirement that space be of constant curvature. Helmholtz's theorem was proven in a more precise and general form by Sophus Lie, and is known as the Helmholtz–Lie theorem. For more details see, e.g., Torretti (1977, chapter 3), or Stein (1977).

17. See, for example, Magnani (2002), Ben-Menachem (2001), Friedman (1999b), or Toretti (1977).

18. See, for example, Friedman (1999a), Coffa (1983), and DiSalle (2002b).

19. This correspondence, and Poincaré's notion of "disguised definition," are discussed by Coffa (1983, 1991, especially pp. 129–34), though with a somewhat different philosophical aim in mind than the present one.

20. This was first pointed out by Friedman (1999b), by which the present discussion has been (obviously) heavily influenced. See also DiSalle (2002b).

21. See, for example, Spivak (1967, volume III) or Bishop and Goldberg (1980).

CHAPTER 4

The origins and significance of relativity theory

General relativity once seemed to be philosophically clear. Physics, according to the logical positivists, had come together with the leading ideas of epistemology, metaphysics, and the foundations of geometry into a single coherent picture – something that had not happened since the philosophy of Kant. In Kant's case, however, what united Newtonian physics, Euclidean geometry, and the critical philosophy was a naive conception of mathematics as a creature of sensible intuition. That conception was, as we saw, overthrown by nineteenth-century ideas: the emergence of non-Euclidean geometry, the rise of conventionalism, and, in general, the separation of formal mathematics from intuition. To the positivists, general relativity was no more or less than the synthesis of these post-Kantian ideas with an empiricist view of science.

The positivists' notion now seems to be as naive as Kant's. But this is not because they were utterly misguided about the philosophical significance of relativity. Rather, it was because they misunderstood the philosophical relations between relativity and what came before it. They could not fully understand the nature of the radical change that Einstein effected, as long as they failed to appreciate the essential philosophical continuities between his theories and those of Newton. They could not see a satisfactory alternative to Kant's theory of the synthetic a priori, as long as they were fixed on the idea of arbitrary convention. Nor could they make much progress on either of these problems as long as they misunderstood the fundamental role that conceptual analysis had played in articulating a reasonable empiricist view of geometry. This chapter, then, does not attempt a complete historical account of all of the developments leading to special and general relativity (which, in any case, others have already done successfully).[1] Instead, it aims at a new account of Einstein's philosophical arguments for special and general relativity, by placing them in their proper context: first, in relation to the philosophical history recounted in the previous chapters, and second, in relation to the general questions about philosophy and space-time physics that this book attempts to address.

4.1 THE PHILOSOPHICAL BACKGROUND TO SPECIAL RELATIVITY

The conception of physical geometry that emerged by the end of the nineteenth century, one might say, resulted from the emergence of the problem of interpretation. Once it was possible to think of a multiplicity of possible geometries as formal structures, it became possible to question the ways in which the formal structures may be applied to experience, and to see the necessity for reflection – and possibly a conventional decision – on which aspects of experience represent aspects of geometrical structure. For Kant, there could be no uninterpreted, purely formal geometry, as the intuitive interpretation was the condition of the possibility of any geometry at all. Even for Helmholtz, geometry was defined as the science of the displacements of rigid bodies, and had no need of interpretation. Conventionalism seemed to be the natural consequence of understanding geometry in far more abstract terms than Helmholtz's or Kant's, as a theory of a much more general class of purely formal structures.

It was not obvious, however, that this new understanding of geometry should have any special implications for physics. A comparable revolution in physics would require, first, some reason to question the principles of classical physics; second, some reason to think that the problems of physics concerned its most general and basic presuppositions, namely its implicit conceptions of geometry, space, and time. That Einstein saw just such a connection, between the difficulties of theoretical physics and its assumptions about the measurement of space and time, was no fortuitous accident. Nor did it arise from an inexplicable flash of insight. To understand it fully, we need to examine Einstein's philosophical engagement with the prevailing physics of his time. This involves taking Einstein, to a certain extent, at his word: he claimed on many occasions that the philosophical arguments with which he introduced his theories corresponded to the ways in which he arrived at the theories himself. The challenge of this approach is to reconstruct his philosophical arguments in a way that makes them seem plausible. If we could do this, we would make some important progress beyond the logical positivists, who represented those arguments rather unconvincingly as applications of simplistic epistemological rules. We might also begin to answer the challenge posed by Kuhn, by showing that Einstein's arguments against the Newtonian views were not circular ones that took his own view for granted. Rather, they began with the spatio-temporal presuppositions of the Newtonian views, and showed how their inadequacies revealed the principles on which an entirely new theory could be built.

It is sometimes said that special relativity overthrew the concepts of absolute space and time. This is only half true; by the time Einstein began his work on electrodynamics, the concept of inertial frame was already widely known, and absolute space was already widely understood to be superfluous (see DiSalle, 1991). Thomson (1884) introduced the notions of "reference-frame" and "reference-dial-traveller," i.e. a spatial frame and a temporal standard relative to which motion may be measured, so that the laws of motion may be stated thus: for any system of particles moving anyhow, there exists a frame and a time-scale with respect to which every acceleration is proportional to and in the direction of an applied force, and every such force belongs to an action–reaction pair. Moreover, any frame in uniform rectilinear motion relative to such a frame is also an inertial frame. Independently, Lange (1885) offered an essentially equivalent conception, the "inertial system" and "inertial time-scale," and Lange's version (and terminology) was more prominently discussed in the German-language literature that Einstein might have read. It was especially emphasized by Mach, in the second (1889) and later editions of *Die Mechanik*.[2]

How much Einstein absorbed of all of these discussions is not clear. It is clear, however, that by 1905 he must have thought it completely uncontroversial that mechanics has no need of absolute space, but needs only "a coordinate system in which the equations of mechanics are valid" (1905, p. 892); by the relativity principle, any system that is in uniform motion relative to such a system is physically equivalent to it. The only question was whether electrodynamics stood in violation of this relativity principle, by treating electromagnetic processes as waves propagating with a definite velocity in a stationary medium, the ether. Central supporters of the ether theory, such as Maxwell, asserted that the relativity principle was still upheld: the velocity of light relative to the ether is, after all, still a relative velocity (Maxwell, 1877, p. 35). Therefore one might maintain the equivalence of inertial frames, while acknowledging that one subset of them happens to represent the rest-frame of a certain physical object whose states determine electromagnetic phenomena. At the same time, Lorentz and his contemporaries confronted the peculiar fact that motion relative to the ether is impossible to detect: the effects of such motion on the relative velocity of light had been calculated, and the Michelson–Morley experiment was sensitive enough to produce the effects of the motion of the Earth through the ether, but no effects were detected (then or since). Lorentz explained this null result by the contraction of all objects, including our measuring apparatus, in proportion to their velocity relative to the ether. For Einstein, however, rest in the ether was tantamount to absolute rest, and the extension of the relativity principle to electrodynamics was therefore an open problem. The apparent indistinguishability of the ether frame meant that the problem was urgent as well.

Einstein made it clear from the beginning that the solution would require some analysis of fundamental concepts. According to the logical positivists, this meant an "epistemological analysis of the concept of time": epistemology requires an account of the empirical meaning of the concept, and physics provides a process – light signaling – that enables us to define the concept in empirical terms. So the revolutionary significance of special relativity lay in this: Newtonian physics had an abstract conception of "absolute" simultaneity, but no physical definition of it; Einstein saw the need for a "coordinative definition" by which simultaneous events could be identified. He therefore supplied the need by introducing an essentially arbitrary stipulation, that when a light signal is propagated from a point A and reflected at B, the time of propagation from A to B is the same as the time from B to A. Thus a vague and "metaphysical" conception of simultaneity is replaced by one that is empirically meaningful, through the act of stipulating what its empirical meaning shall be.

It should not be difficult, then, to understand why Einstein's account of simultaneity seemed to encourage a verificationist account of meaning. Evidently he had given an analysis of the concept that was, at the same time, a rule for verifying that two events are in fact simultaneous. Moreover, Einstein's own language sometimes suggested that he viewed the matter in just this light. His 1905 paper asserts that "a mathematical description [of the motion of a material point] has no physical meaning unless we are quite clear as to what we understand by 'time,'" and goes on to consider possible empirical methods of synchronization. He elaborates in his popular account of relativity: "The concept of simultaneity does not exist for the physicist until he has the possibility of discovering whether it is fulfilled in an actual case. We thus require a definition of simultaneity such that this definition supplies us with a method by means of which, in the present case, he can decide by experiment whether the two lightning-strokes occurred simultaneously" (Einstein, 1917, p. 22). Reichenbach drew the lesson most explicitly: "The physicist who wanted to understand the Michelson experiment had to commit himself to a philosophy for which the meaning of a statement is reducible to its verifiability, that is, he had to adopt the verifiability theory of meaning if he wanted to escape a maze of ambiguous questions and gratuitous complications" (Reichenbach, 1949, pp. 290–1).

If the concept of simultaneity is to be defined by its means of verification, the question arises just how the means ought to be determined. Einstein himself often suggested that it was a matter of conventional choice. In 1905 he asserts that a "common time" for different observers can be defined only if we "establish *by definition* that the 'time' required by light to travel from A to B equals the 'time' it requires to travel from B to A" (Einstein, 1905, p. 894). But he gives no explicit justification for the use of light signals in particular. And in later remarks, he speaks as if the isotropy of light propagation, and its use in time measurement, is fixed by an arbitrary stipulation. In his popular exposition of his work (1917), he considers a possible objection to his principle: how can we test the hypothesis that the speed of light is isotropic, unless we already have a way of measuring time? The answer is that the principle is only a definition. "Only *one* requirement is to be set for the definition of simultaneity: that in every real case it provide an empirical decision about whether the concept to be defined applies or not"; that light takes the same amount of time to travel in both directions "is neither a supposition nor a hypothesis, but a stipulation that I can make according to my own free discretion, in order to achieve a definition of simultaneity" (Einstein, 1917, p. 15). In his Princeton lectures (1922), he raises the question why light propagation should play such a central role in his theory, and gives no more answer than that "It is immaterial what kind of processes one chooses for such a definition of time," except that it is "advantageous . . . to choose only those processes concerning which we know something certain" (Einstein, 1922, pp. 28–9).

Remarks like these may be taken to suggest that, after all, the positivists' view of the origins of special relativity was very near the truth. Whatever attempt we might make to justify the theory on inductive grounds, and however well such an attempt might succeed – in order to show that it is simply better confirmed than its predecessors – Einstein saw himself as investigating the meanings of fundamental concepts, and as linking their meanings to the empirical procedures that determine their application. Moreover, he noted the element of arbitrary convention in the choice of criteria. It is not surprising that he should have done so. After all, he claimed to have been profoundly influenced by Poincaré's writings. Nor is it surprising that his account of his own reasoning should resemble that of the positivists, since that reasoning was just what their account of science was attempting to capture. In short, it is clear that in Einstein's mind, the problems of electrodynamics were connected with the a-priori principles of physics, that is, with the concepts of spatial and temporal measurement from which the empirical study of physics must begin. He suggests as much in his 1905 paper: "The theory that is to be developed rests – like all electrodynamics – on the kinematics of the rigid body, since the assertions of any such theory concern the relationships between rigid bodies (systems of coordinates), clocks, and electromagnetic processes. Insufficient consideration of this circumstance lies at the root of the difficulties which the electrodynamics of moving bodies presently has to struggle" (Einstein, 1905, p. 892). The question is whether, in interpreting Einstein in this way, we are compelled to accept the arbitrariness of his starting point. Einstein deduced the Lorentz transformations from the invariance of the velocity of light, but, of course, the argument can go in the opposite direction: instead of explaining the seeming contraction by the invariance of the velocity of light, we can explain the seeming invariance of the velocity of light by the real contraction of our measuring devices. The logical relation between the two principles cannot by itself determine that one or the other deserves to be regarded as more fundamental. Lorentz's theory, no less than Einstein's, might be regarded as a "natural" (as opposed to an artificial or ad hoc) way to explain the Michelson–Morley results without resorting to arbitrary hypotheses. As Lorentz himself said in describing the aim of his theory,

Surely this course of inventing special hypotheses for each new experimental result is somewhat artificial. It would be more satisfactory if it were possible to show by means of certain fundamental assumptions and without neglecting terms of one order of magnitude or another, that many electromagnetic actions are entirely independent of the motion of the system . . . The only restriction as regards the velocity will be that it be less than that of light. (Lorentz, 1904, p. 13)

Such a remark suggests that Lorentz had some of the same methodological concerns that we typically attribute to Einstein, and therefore casts doubt on the notion that Einstein's approach was superior on general methodological principles. The essential argument against Lorentz lies elsewhere, in the analysis of simultaneity.

4.2 EINSTEIN'S ANALYSIS OF SIMULTANEITY

The starting point of Einstein's argument is well known: "Maxwell's electrodynamics . . . in its application to moving bodies, leads to asymmetries which do not appear to be attached to the phenomena" (Einstein, 1905, p. 891). For example, a phenomenon that depends only on the relative motion of a conductor and a magnet – the production of an electric current – is represented by the theory in two completely different ways, depending on whether the conductor or the magnet is taken to be at rest. Assuming that the conductor is at rest, there is an electric field in the vicinity of the magnet; assuming that the magnet is at rest, there is an electromotive force in the conductor. But the measurable magnitude – the current – is the same in both cases, as long as the relative motions are the same. If the Michelson–Morley experiment had no great influence on Einstein's thinking, it is doubtless because, as Einstein himself suggests, it was only an additional example of the kind of empirical symmetry that had concerned him already on independent grounds.

At first glance, this problem may appear to be no more or less serious than the problem of absolute space in Newton's theory, and the analogous asymmetry between absolute space and uniformly moving frames. By the same token it may seem as if eliminating the asymmetry were no more difficult or serious than asserting the equivalence of all Newtonian inertial frames and thereby eliminating absolute space. For these reasons it might be tempting to assert a straightforward methodological justification for special relativity, as merely eliminating a theoretical distinction that makes no difference. Yet the asymmetries that Einstein notes are considerably more serious. On the one hand, the theoretical asymmetry is a kind of ontological asymmetry, in which, depending on what is taken to be at rest, a different sort of field is said to exist. One might see a similar ontological asymmetry in the case of absolute space. But there it was clear even to Newton that, however absolute space may be bound up with his general metaphysical picture, it could be completely disregarded in our conception of the physical entities and processes at work in a Newtonian world: the ontology of bodies moving under the influence of accelerative forces does not require any distinction between uniform motion and rest. So the elimination of the distinction, and of absolute space, in no way disturbed the fundamental physical concepts of Newton's theory. On the contrary, it merely brought the theory of space and time into complete harmony with those concepts. In the electrodynamical case, the asymmetry is essential to the conception of electromagnetic forces as mediated by waves in the ether. Eliminating that asymmetry seemed, as Einstein noted, to involve us in a contradiction, namely between the Galilean principle of relativity (the "relativity postulate") and the principle that the velocity of light is independent of the motion of the source (the "light postulate"). In retrospect, we have been convinced by Einstein that this is only an apparent contradiction. But it is entirely genuine if we presuppose a framework of concepts about space, time, and electrodynamics that seemed perfectly reasonable to presuppose at the time. So Einstein faced a twofold task of philosophical analysis: first, to determine precisely what it was, in the accepted framework of assumptions, that the contradiction rested upon; and only then, to discern the basis on which an alternative could be constructed. The alternative, too, had a twofold burden: not only to construct a framework in which the contradiction would not arise, but also to show that the fundamental concepts of this framework were well-defined in a way that the previous concepts were not. For, without an argument for the second point, Einstein could only make the subjective argument, that his hypothesis could explain the same phenomena as Lorentz's in a more "natural" way. To his own mind, at least, he was doing something more than this.

What seems natural within the Lorentz theory is that velocity is relative; even the "true" velocity of light is only its velocity relative to a particular material system. Thus the invariance of the velocity of light makes no sense, and the apparent invariance, as revealed by the Michelson–Morley experiment, is a phenomenon to be explained. In fact, within the broader Newtonian framework – methodological as well as spatio-temporal – such a phenomenon *must* be interpreted more or less along Lorentz's lines, i.e., as providing information about the presence of some distorting force or other. Hence Lorentz's quite plausible comment on the contraction hypothesis, that, "we shall have to admit that it is by no means far-fetched, as soon as we assume that molecular forces are also transmitted through the ether, like the electric and magnetic forces of which we are able at the present time to make this assertion definitely . . . Now, since the form and dimension of a solid body are ultimately conditioned by the intensity of molecular actions, there cannot fail to be a change of dimensions as well" (Lorentz 1895, p. 6).

Einstein eventually argues that this entire project of explanation rests on questionable grounds, namely, on assumptions about space and time that have not been sufficiently examined. But this is not, as was afterwards claimed by the positivists, because the Newtonian framework altogether lacks a coordinative definition of simultaneity. Rather, it is that the coordinative definitions that do exist have a very problematic status. His argument starts by taking for granted an inertial system, or "a system of coordinates in which the equations of mechanics hold good (i.e. to first approximation)" (Einstein, 1905, p. 892). But this starting point is quickly revealed to be ironic. The equations of mechanics concern the motions of a material point relative to such a coordinate system. But before we can describe the motions of a material point, we need to define what we mean by time. And if we cannot take for granted the kinematical description of motion yet, then we cannot take for granted the inertial coordinate system either; both stand in need of a definition of time. This is something quite different from the problem of having to accept the equations of motion "to first approximation," with the expectation of small revisions to the equations as the analysis proceeds. The problem is, rather, that the starting point simply cannot be taken at face value, because we have not defined its basic terms. Hence Einstein's attack on the notion of simultaneity is essentially a dialectical one.

This account of the problem may not seem plausible at first. In principle – and in spite of the criticisms of the positivists – there is a perfectly good definition of a Newtonian inertial frame, and a perfectly good definition of simultaneity. As we have already seen, Newton's conception of simultaneity is instantiated by the instantaneous propagation of gravitational force; in principle, it ought to be possible to know something immediately about spatially distant states of affairs. That there is no *practical* application of this criterion does not, by itself, justify the claim that absolute simultaneity had no empirical meaning, and that Einstein had provided a coordinative definition of simultaneity where none had existed before. It would be reasonable to suspect, indeed, that light signals could provide a kind of stand-in for an infinitely fast signal, as long as we can take into account the travel time of the light signals, or of any signal whose velocity is known and reasonably constant – though electromagnetic waves apparently stand alone in this regard. This situation, and its significance for the classical view of space and time – or, more precisely, its seeming lack of significance – was articulated very clearly by James Thomson. He appears to have been the first to remark that the measurement of distance involves

the difficulty as to imperfection of our means of ascertaining or specifying, or clearly idealizing, simultaneity at distant places. For this we do commonly use signals by sound, by light, by electricity, by connecting wires or bars, and by various other means. The time required in the transmission of the signal involves an imperfection in human powers of ascertaining simultaneity of occurrences at distant places. It seems, however, probably not to involve any difficulty of idealizing or imagining the existence of simultaneity. Probably it may not be felt to involve any difficulty comparable to that of attempting to form a distinct notion of identity of place at successive times in unmarked space. (Thomson, 1884, p. 380)

This passage states remarkably clearly what is at stake with the notion of simultaneity: the very idea of a kinematical frame of reference. For even the measurement of spatial distance presupposes the ability to determine simultaneous events. The passage also makes clear how natural it was, at the time, to assume that the dependence of simultaneity on signal propagation poses no special problem of principle. As late as 1910, Simon Newcombe saw it in purely practical terms: "Were it possible by any system of signals to compare with absolute precision the times at two different stations, the speed [of light] could be determined by finding how long was required for light to pass from one station to another at the greatest visible distance. But this is impracticable, because no natural agent is under our control by which a signal could be communicated with a velocity greater than that of light" (Newcombe, 1910, p. 623). In other words, this problem of dependence was noted long before it appeared to give rise to a problem of *relativity*.

It is quite unfair, then, to suppose that Newtonian physicists did not appreciate the role of the concept of simultaneity within the Newtonian system, nor the need to attach some physical meaning to it. Nonetheless, in order to understand the force of Einstein's challenge, we must understand on what precarious ground the Newtonian concept of simultaneity really stood – and why the entire conceptual system was thereby at stake. Again, the best possible instantiation of absolute simultaneity was universal gravitation. From the law of velocity-addition, it is true, one could deduce the possibility of particles accelerated to arbitrarily high velocities. But the only known case of instantaneous propagation, even in theory, is gravitation. Yet gravitational signaling is, obviously, not a constitutive principle of the Newtonian framework; in fact it was discovered by "reasoning from phenomena," according to Newton, a kind of reasoning that necessarily takes the Newtonian spatio-temporal framework for granted. In particular, interpreting the Solar System as a dynamical system already presupposes that we can measure the relative distances of the planets, their satellites, and the Sun, as well as their angular positions relative to the fixed stars. All of this presupposes the theoretical possibility, at least, of determining simultaneous events. Our ability to determine all of these circumstances, however, depends on a physical process – light signaling – that is not only finitely propagating, but also essentially *extraneous* to the Newtonian theory. The latter is not necessarily a problem in itself; on the contrary, it might be seen as an advantage to have an observational technique that does not rely upon the theory being tested. But it does reveal rather starkly how the Newtonian approach to celestial mechanics begins with no criterion of simultaneity other than the "intuitive" criterion of visual perception, namely light signaling.

With a physical picture of the Solar System in hand, the role of light in establishing the background framework may be set aside. Indeed, the framework now permits the empirical study of light propagation as a process unfolding within this framework. A familiar example is the estimate of the speed of light from the delays in the eclipses of Jupiter's moons; by observing how the timing of the delays depends on the relative positions of the Earth and Jupiter, Ole Roemer was able to calculate the time of light propagation from Jupiter and to reach a remarkably accurate result. Such a calculation begins, in effect, from the assumption that certain events are objectively simultaneous even though they appear to us successively. It assumes, moreover, that the relative *velocity* of Jupiter and the Earth is small enough (relative to the speed of light) to ignore. In other words, the visual criterion of simultaneity is revealed, in this development, to be only a kind of stand-in for an instantaneously propagating signal, the kind of signal that gravity represents but does not provide for any practical purpose. The criterion can be assumed to be perfectly adequate for a system of bodies that are relatively at rest (when we correct for different distances), and approximately accurate for systems where the relative velocities are negligible. And one could reasonably expect that, if the relative velocities became too great to be neglected, the criterion could still be applied by applying the Newtonian rule of velocity-addition.

Absolute simultaneity, then, is an *abstraction* from the empirical practice of determining simultaneity by signals, an abstraction that involves either neglecting the time involved in signaling, when it is sufficiently small, or extrapolating that any measurable time delay may be accounted for. This process of extrapolation exemplifies Newton's view of absolute time as a measurable quantity – that is, his construction of absolute simultaneity and equal time intervals as concepts to which we can expect to arrive at increasingly good approximations. It is precisely this expectation that turns out to be so precarious. If the velocity of light turns out not to obey the classical addition law, then this entire line of approximative reasoning is fatally undermined. If that is true, then the other supposed instantiation of absolute simultaneity – universal gravitation – becomes essentially *hypothetical.* This does not mean that gravity no longer instantiates absolute simultaneity, or that the latter is no longer a meaningful concept. It does mean, however, that the concept is no longer integrated systematically with our account of its physical measurement; the gravitational definition now stands on its own, no longer connected with the empirical definition by a series of approximative steps.

This is not a conclusion to which we are forced by the Michelson–Morley experiment. Indeed, to someone intent on interpreting that experiment, such as Lorentz, simultaneity might not appear to be in question at all. We noted that the unexpected null result of the experiment, from Lorentz's point of view, must be seen as providing information about the influence of some unknown factors. Lorentz himself suggested that the results reveal an effect of motion through the ether on intermolecular forces, and that this effect produces the contraction of moving objects. In fact this is probably the only reasonable approach to take if the underlying conceptions of simultaneity, length, and time are not questioned. Einstein, however, was not primarily concerned with interpreting the Michelson–Morley experiment. Rather, he was concerned with understanding why *no* electrodynamical processes seem to distinguish between uniform motion and rest, even in cases where the theoretical treatment of them explicitly appeals to absolute velocities. If this phenomenon is contradictory, then he had to expose exactly what it contradicts. Therefore he saw the need to examine the assumptions about spatial and temporal measurement from which the contradiction arises. It is contradictory for the laws of electrodynamics to respect the relativity principle, if the relativity group of physics is the Galilean group, whose invariants include mass, acceleration, length, and time – and therefore simultaneity – but cannot include a velocity. But if the underlying spatio-temporal concepts are ill-defined, then the relativity principle itself has to be defined in completely different terms.

This is why the "Kinematical Part" of Einstein's 1905 paper begins just as it does, by arguing that the concept of a Galilean coordinate system, taken for granted in the Galilean principle of relativity, cannot be accepted without an adequate definition of time. It is in the context of this argument that the definition of simultaneity by light signals emerges, not merely as convenient, but as uniquely satisfying the requirements imposed by the nature of the problem. Einstein's series of proposed definitions – including seemingly naïve references to "the hands of my watch" and the like – have an apparently "operationalist" aspect that has often been remarked upon. But Einstein's discussion does not really invoke a truly operationalist view. The requirements of the problem are theoretical as well as practical, and the assumptions that eventually go into the definition involve both empirical facts and the theoretical structure of Maxwell's electrodynamics. In fact Einstein proceeds through a brief series of possible operations for determining simultaneity, each time showing that theoretical requirements make the proposed definition inadequate, until he arrives at the one that is satisfactory. It is not really an exaggeration, indeed, to say that Einstein's analysis aims to arrive at a definition of "absolute" simultaneity, and in something like the sense intended by Newton. This is not to suggest that he imposes in advance the Newtonian condition that which events are simultaneous shall not depend on the state of motion of the observer. But he seeks a *criterion* of simultaneity that is independent of position and motion, that has a foundation in physical laws that are independent of any observer, and that makes simultaneity a symmetric and transitive relation. It is a kind of Socratic irony that the criterion of simultaneity that eventually satisfies these requirements, and so deserves to be called "absolute" in this special sense, finally makes the relation of simultaneity a relative one.

Einstein begins by proposing two practical procedures, each of which supplies an "operational" criterion of simultaneity: to define "time" by "the position of the small hand of my watch," and to coordinate the time of every event with a watch at a fixed location, by the time at which a light signal from each event reaches the watch. The first obviously fails to meet the requirement of defining simultaneity for distant events; the second meets that requirement, but "has the disadvantage that it is not independent of the standpoint of the observer" (Einstein, 1905, p. 893). Thus each of these proposed criteria fails the theoretical requirements in some way. But then Einstein introduces his final criterion: a common time for points A and B can be defined when we "establish *by definition* that the 'time' required by light to travel from A to B equals the 'time' it requires to travel from B to A" (Einstein, 1905, p. 894). Placing "time" in quotation marks emphasizes the fact that we are not to associate some pre-theoretical or intuitive meaning with the concept; again, if there were a principled way to do that, Einstein's statement could be an empirical claim rather than a definition. But this evidently is a definition, and one that evidently does satisfy Einstein's requirements. It exploits the invariance of Maxwell's equations as a foundation for an objective theoretical concept.

At this point in the argument, the true significance of the empirical facts, including the null results of the Michelson–Morely experiment, need not be assumed to be known. Whatever its true significance – it is not yet ruled out that this is the effect of a universal contraction of bodies moving through the ether – the apparent invariance of the speed of light guarantees that light signaling will satisfy Einstein's requirements for an adequate definition of simultaneity. Whether it uniquely satisfies them is another question altogether, and one that could never be answered conclusively; no argument of Einstein's could prove that there is nothing in the world that travels faster than light, or no more appropriate signal for determining simultaneity. But it is clear, at least, that the general demand that Einstein articulated later – "that in every real case it provide an empirical decision about whether the concept to be defined applies or not" – is satisfied, and, moreover, that its satisfaction involves much more than specifying an operation. In fact it involves specifying the place of the definition within our systematic knowledge of physical laws. The Newtonian definition, as we saw, is deeply embedded in the entire framework of Newtonian mechanics and gravitation theory. But its connection with empirical criteria was always somewhat tenuous, and, again, dependent on assumptions extraneous to the Newtonian program. Once that connection was called into question, by unexpected facts about light propagation, Einstein's definition stands as an implicit challenge to the Newtonian conception of time: what physical principles can define simultaneity for observers in relative motion?

So baldly stated, and taken by itself, this challenge might have appeared to invoke the positivists' verificationist account of meaning. Now that we have considered the conceptual analysis leading to it, and its theoretical dimension, we can see that it is considerably more complicated and interesting than the positivists took it to be. It cannot be characterized as "an epistemological analysis of time" in some general sense, demanding an empirical account of "what we mean by simultaneity." In fact Einstein's analysis does not question what we ordinarily mean by time, since the most common way of identifying simultaneous events is by observing them at the same time; once the velocity of light is acknowledged to be finite, it is a minor matter of error-correction to adjust this criterion for the different distances at which events take place. What Einstein does question is the relation of this common-sense notion to our systematic knowledge of physics, and the peculiar status it acquires in the contemporary state of physics. On the one hand, because of the apparent failure of velocity-addition, Newtonian physics no longer provides a theoretical context that connects the common-sense criterion with the "absolute" conception. On the other hand, and for the same reason, the invariance properties of electrodynamics place this criterion in a peculiar "absolute" position – standing on its own, rather than as a local or slow-moving approximation to some more fundamental criterion. In other words, we could say that Einstein's analysis respected the intuitive *criterion* of simultaneity, but recognized for the first time that it had been disconnected from the intuitive *theory* of simultaneity. On the intuitive theory, it was supposed to be an objective fact whether any two events happened at the same moment, or successively (and in which order). But it turned out to be only a shaky theoretical edifice, after all, that had connected ordinary judgments of simultaneity with the absolute distinction of past, present, and future.

Special relativity can be thought of as the theory that the fundamental symmetry group of space-time is not given by the Galilean transformations, but by the Lorentz transformations

$$x' = \frac{x - vt}{\sqrt{1 - v^2/c^2}}$$
$$y' = y$$
$$z' = z$$
$$t' = \frac{t - vx/c^2}{\sqrt{1 - v^2/c^2}}$$

In 1908, Minkowski developed what is now the most familiar representation of the special theory, as a four-dimensional geometry with an indefinite metric – that is, as the "absolute" space-time structure underlying the relativity of space and time individually. As in Newtonian space-time (see Stein, 1967), "the world" is a four-dimensional affine space, and inertial trajectories are the geodesics of this space. But there is no invariant way to decompose it into sets of simultaneous events. The invariance of the velocity of light corresponds to the invariant metrical interval of the four-dimensional pseudo-Euclidean space; simultaneous events relative to an inertial observer's frame of reference are represented by the hyperplane orthogonal to the observer's worldline. It follows that there is no objective division of the world, at any moment, into past, present, and future – indeed, no moment at all, except as a given point on the worldline of a given observer. Instead there is, at any point, a "future light-cone" whose surface represents all the events accessible from that point by a light signal, and whose interior represents all the events accessible by any material particle moving slower than light, along with a corresponding "past light-cone" of events *from* which the given point may be reached. (See Figure 5.) The set of events at a given interval from a given point is not, as in Euclidean space, a sphere with that point at its center, but a three-dimensional "hyperboloid of revolution" – except that, because of the peculiar invariance of light propagation, all events on the light-cone lie at a null interval from its origin, and the path of a light ray is orthogonal to itself. All the peculiar features of special relativity, regarding the comparison of length and time in different inertial frames, can be easily derived from the features of this geometrical structure.[3]

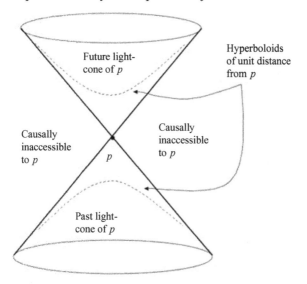

Figure 5. The causal structure of Minkowski space-time: space-time at any point *p* is divided by the "light-cone" into regions that are accessible or inaccessible by any causal influence (i.e. any signal traveling at or less than the speed of light).

For our discussion, the interest of this structure lies in Minkowski's account of its origins and significance. We might simply regard it as the application of a convenient, pre-existing mathematical formalism to a new physical theory – an instance of what Minkowski called the typical "staircase wit" of mathematics, easily capturing the mathematical essence of a new theory after physics has done all the difficult work (Minkowski, 1909, p. 105). Once we see that the invariant quantity in Einstein's theory is a spatio-temporal quantity, the speed of light in vacuo, *c*, we can represent the rate, time, and distance in a given coordinate system by

$$c^2 t^2 = x^2 + y^2 + z^2$$

and in a relatively moving frame by

$$c^2 t'^2 = x'^2 + y'^2 + z'^2$$

Then it is evident that we can represent the invariance of the speed of light by the expression

$$c^2 t^2 - x^2 - y^2 - z^2 = c^2 t'^2 - x'^2 - y'^2$$

To Minkowski, thoroughly schooled in Klein's approach to geometry as a theory of structure and automorphisms, it might have seemed obvious that either side of this last expression "looks like" the expression for a metric in four dimensions; this is only slightly obscured by the presence of c, which may be trivially set at unity, and the fact that the coefficients for the spatial and temporal parts of the expression are opposite in sign. We can even make it look exactly like a Euclidean metric in four dimensions, $x^2 + y^2 + z^2 + t^2$, by setting $c = 1$ and $t = -ict$. But these are merely mathematical facts. What Minkowski emphasized, above and beyond these facts, was that special relativity had arisen from a conceptual analysis of the role of time in electrodynamics, and that the new space-time structure arose from a further conceptual analysis of what Einstein had revealed about the nature of time.

As we noted already, Einstein's theory can be regarded as nothing more than an alternative hypothesis to Lorentz's, starting by "raising to the status of a postulate" what Lorentz hoped to explain away as a mere appearance. One might then, as many have with good reason, defend Einstein's view as more simple, more "natural," or in better accord with some other methodological canon. But for Minkowski, such an approach would blur a fundamental distinction between the two theories. Einstein has shown, Minkowski writes, that the relativity postulate "is not an artificial hypothesis, but rather a novel understanding of the time-concept that is forced upon us by the appearances" (Minkowski, 1908, p. 56). It is Lorentz's theory that involves a hypothesis, in order to explain the differences of local time for electrons in relative motion; Einstein's theory arises from the analysis of this difference of local time in relation to the concept of simultaneity, and shows that this empirical fact reveals something about the nature of time in general.

Lorentz called the t' combination of x and t the local time of the uniformly moving electron, and applied a physical construction of this concept, for the better understanding of the hypothesis of contraction. But to have recognized clearly that the time of the one electron is just as good as that of the other, that is, that t and t' are to be treated equally, was first the merit of A. Einstein. (Minkowski, 1909, p. 107)

The recognition of this equivalence, and of its fundamental character, is what makes the analogy between the four-dimensional Minkowski structure and spatial geometry something more than a mere formal analogy. The group of spatial displacements defines the structure of space in the obvious way that Poincaré had articulated, and, as long as this structure could be considered entirely independently of dynamics, the spatial structure could be regarded as the basis of physical geometry. But that independence assumed the independence of space from time, as implied in the assumption of absolute simultaneity. By recognizing the relativity of time, Einstein had raised the *spatio-temporal* displacements, the Lorentz transformations, to a privileged status that the spatial displacements could no longer legitimately claim – because, unlike the Galilean transformations, the Lorentz transformations implied that spatial geometry was no longer a fundamental invariant. The physically objective quantities must be expressed as the invariants of a four-dimensional structure.

[We] are bound to admit that it is in four dimensions that the relations being considered here first reveal their inner essence in full simplicity, but on a three-dimensional space previously imposed upon us they cast only an extremely complicated projection. (Minkowski, 1909, p. 110)

Einstein's 1905 paper only states the true spatio-temporal relations in a form bound by the limits of spatial intuition.

This is the background for one of Minkowski's most familiar remarks, that "the relativity-postulate" is a "feeble" and inappropriate name for Einstein's idea:

Insofar as the postulate comes to mean that only the four-dimensional world in space and time is given by the phenomena, but that the projection in time and space may still be undertaken with a certain freedom, I would rather give this claim the name, the *postulate of the absolute world* (or, briefly, the world-postulate). (Minkowski, 1909, p. 107)

It is perhaps too easy, and certainly not uncommon, to overstate the significance of this "world-postulate" and what Minkowski meant by it. Einstein had stated a principle of relativity for electrodynamics, and derived the transformations that relate the various perspectives of possible observers. If Minkowski described the geometry of the unified "absolute world" that underlies all of these relative perspectives and this group of transformations, should we say that he has explained the Einsteinian relations? To do so would be, at the very least, a loose way of speaking. Minkowski space-time is not presented as a deeper sort of reality underlying the phenomena described by Einstein, explaining them as, say, the kinetic theory of gases explains the phenomena described by the ideal gas law. A more careful assertion would be one that acknowledged, at least, that a very different kind of explanation is at work. Minkowski space-time has been described as providing a "structural explanation" (see Hughes, 1987): the phenomena are to be explained by the fact that the world is a model of a certain structure. But then in this case, the explanation would have to be that the world conforms to Einstein's relativity principle because it is a model of Minkowski space-time. Then the seemingly explanatory character of such a statement is quite misleading. For Minkowski's structure is in fact only the expression of Einstein's theory in other terms, namely the expression of a "three plus one"-dimensional theory in four-dimensional terms. In order to explain the physical meaning of the Minkowski structure, we could do little more than make the same assertion in reverse: the world has that structure because the laws of electromagnetic propagation are the fundamental invariants, that is, because Einstein's theory is true. Consider, once again, the example of ordinary spatial geometry: if we know that there is a group of rigid displacements that preserves the Pythagorean interval, do we really explain this by saying that the world is a model of Euclidean space? Clearly not: to say that such relations exist among real objects is only to restate the claim that real space is Euclidean. As we saw in Chapter 3, understanding the content of geometrical claims in this way was the great philosophical achievement of Helmholtz and Poincaré, and the beginning of the modern understanding of physical geometry in general.

Yet, from another perspective, just this trivial non-explanation is of the highest importance. First of all, it is a fairly obvious point that the mere representation of Einstein's theory in a different form is not without physical significance. Alternative formulations of the same theory, though they are by hypothesis equivalent, are not equivalent for all intents and purposes: for the purpose of seeing how the theory may relate to other theories, or even of pursuing entirely new theoretical directions, it is clear that alternative formulations may be suggestive or fruitful in quite different ways. The history of physics offers many familiar examples of this, and it is a familiar enough topic of philosophical discussion; one obvious example is Minkowski space-time itself, for it is hard to imagine how general relativity could have developed from special relativity had the latter not been recast in four-dimensional terms. Also crucial, for the eventual development of general relativity, was the reformulation of Newton's law of gravitation in Poisson's equation for the gravitational potential. The physical significance of both of these reformulations first came to light with the representation of gravity as space-time curvature. We can certainly think of this representation as an explanation of a sort, provided that we don't confuse understanding a theory from a different viewpoint with deriving it from some deeper ontological ground.

For our purposes, however, Minkowski's account does have a deeper though perhaps less obvious significance. For it reveals just how simple and direct is the connection between the structures of space and time and the assumptions we make about physics. The claim at the heart of Minkowski's analysis is, at the same time, extremely far-reaching and extremely modest: it is the claim that a world in which special relativity is true, simply, *is* a world with a particular space-time structure. This is something completely different from the superficially similar claim of Poincaré, that Lorentz's theory can be represented by the group of isometries of a four-dimensional pseudo-Euclidean space (Poincaré, 1905). The difference is analogous to the difference between Lorentz's account of the Lorentz transformations, as expressing the relations between moving and resting electrodynamical systems, and Einstein's account of them as expressing a fundamental symmetry of the laws of physics. In Poincaré's analysis, the spatio-temporal structure expressed by the Lorentz group is not "the structure of space and time"; the structure of space and time must already be determined – by conventional choice – before the analysis of electrodynamics can even begin. A dialectical argument like Einstein's, in which electrodynamics provides a critique of the assumptions about space and time on which electrodynamics itself had been built, was, as we have seen, beyond Poincaré's conception. For Minkowski, however, it was Einstein's argument that revealed the origin and significance of this four-dimensional structure. It represents not merely a convenient way to think about electrodynamics, but how we *must* think about it, in light of Einstein's analysis of simultaneity; what we have learned from Einstein is that this structure represents what we *actually know* about space and time. The connection between the space-time structure and our knowledge of dynamics is just as direct and immediate as that between the structure of Euclidean space and our knowledge of spatial displacements, even though it is much more remote from intuition. In sum, the "postulate of the absolute world" is not the explanation for what Einstein had regarded as merely relative; the world-postulate is simply a better name than "theory of relativity" for what Einstein's theory *actually says*.

It might appear, then, that the theory of space-time as introduced by Minkowski is somewhat superficial, in comparison to the deep ontological questions that we are tempted to ask about it. This peculiar superficiality of space-time theory is something that Einstein came close to articulating, when he designated special and general relativity as "principle theories." Most physical theories, he explained, are "constructive": "They attempt to build up a picture of the more complex phenomena out of the materials of a relatively simple formal scheme," as the kinetic theory of gases explains their gross behavior by "the hypothesis of molecular motion." Principle theories, in contrast, begin from elements that are "not hypothetically constructed but empirically discovered ones, general characteristics of natural processes, principles that give rise to mathematically formulated criteria which the separate processes or the theoretical representations of them have to satisfy" (Einstein, 1919, p. 228). In other words, it is the constructive theories that seek to explain phenomenal regularities by appeal to underlying entities, mechanisms, or processes that are hidden from view. Therefore they cannot avoid being at least partly hypothetical. Principle theories, meanwhile, do not postulate something behind the phenomena, but merely describe the structural constraints that the phenomena actually do exhibit. Lorentz's theory, for example, takes the Lorentz symmetry as a phenomenon to be explained by a constructive theory, namely the theory of the Lorentz contraction, which must ultimately have some molecular basis along the lines suggested earlier. Special relativity, by contrast, identifies the Lorentz symmetry as a general structural constraint that is obeyed by all physical interactions, and that itself cannot be explained in the same sense – it is, in short, a defining principle for an explanatory framework, within which it would make no sense to demand such an explanation for the framework itself. This is why the attempt to explain it must always appear to be circular. Suppose that we purport to explain the constancy of the velocity of light by the claim that space-time has the Minkowski structure; how, then, do we explicate the meaning of the claim, except by repeating that the symmetry group of space-time is one in which the velocity of light is invariant?

It is not surprising that this should be the case. It is only the four-dimensional version of the circularity in defining inertial frames in special relativity: light propagates with the same velocity in any inertial frame, but inertial frames are (by definition) those in which light propagates with the same velocity. In this case as in the Newtonian case, the circle does not indicate a logical defect, but only the fact that we are dealing with a definition, or a principle of interpretation. In the absence of any means of determining simultaneity independently of light signals, we have no means of setting out a spatial frame of reference, or of measuring time intervals in some objective way. So, as Einstein's analysis had shown, our theoretical principles regarding the propagation of light are playing a constitutive role in our spatial and temporal measurement that, for various reasons, we had always been able to overlook or at least under-emphasize. Thus the propagation of light is not something that we can objectively measure within a privileged frame, but part of the interpretation of our concept of an inertial frame. Its role is no different from the role that force and acceleration had played in the definition of a Newtonian inertial frame, except that in that case simultaneity, length, and time had been taken for granted as something entirely independent of the physical principles being invoked. In each case, physical principles determine an interpretation for the concept of inertial frame, which otherwise has no physical content at all. It was therefore somewhat careless of Einstein to say that principle theories are "empirically discovered, general characteristics of natural processes" – as if the principles of special relativity, for example, were merely inductive generalizations. It is true that Einstein regarded the symmetries of electrodynamics as "inherent in the phenomena." But those symmetries by themselves do not constitute special relativity; rather it was Einstein's recognition that they offer the only reasonable definition of an inertial frame. In such a manner, principle theories impose an interpretation upon nature that the search for constructive theories can then take for granted. The reluctance to acknowledge this, naturally felt by the scientific empiricist, stems from the belief that such an imposition must be arbitrary; the only remedy for that is an appreciation of the reasoning, clearly exhibited by Minkowski and Einstein, that draws such an interpretation from the empirical facts.

Einstein never developed his typology of theories beyond a general sketch, and its connection with theories of space and time was never made very clear. Recently, Flores (1999) suggested a way of sharpening the distinction, by considering the peculiar functional relation that exists between principle and constructive theories: principle theories provide a framework for asking empirical questions about physical interactions in general; a constructive theory is developed under the general constraints of a given framework, which permit arguments from empirical evidence about what kinds of physical interaction are at work. For this reason Flores proposed calling them "framework theories" and "interaction theories." This way of making the distinction highlights something that was less than clear to Einstein, namely, that what distinguishes special and general relativity as principle theories is not merely a characteristic of a "theory of theories" or "second-level" constraint on how particular laws are to be formulated (see Earman, 1989, p. 155). In fact it is a general characteristic of theories of space and time. This includes Newton's own theory: Newtonian space-time, as characterized by the Galilean symmetry group, does indeed define a set of general constraints that all physical laws must obey; it thus defines a general framework for inquiry, that is, a framework within which we can investigate forces of nature such as gravity. In just this way Newton himself had inferred characteristics of gravity, and its identity with interplanetary attraction, from the characteristics of the planetary motions.

By not fully appreciating this distinction, Einstein missed a crucial aspect of Newton's theory and its relation to special general relativity. Newton's theory is not, any more than special relativity, a hypothesis about some reality underlying the phenomena. Just *because* it is a theory of space and time, it shares with Einstein's theories the characteristic of being a framework for the interpretation of phenomena, not a kind of mechanism or hypothesis to explain them as Einstein had argued in 1916. Indeed, Newton expressed his own clear recognition of this fact in his distinction between "active" and "passive" principles (1704 [1952], pp. 397–401). The passive principles are the laws of motion: they define the general characteristics of mass and force, and so define general constraints on the form that any physical interaction may take. Taking these principles for granted, along with their consequences for all possible systems (the propositions that Newton derives from the laws in Book I), we can then discover the active principles by empirical reasoning. For these are just the forces of nature, which explain how, within the constraints imposed by the laws, particular systems of particles (like bodies bound by their cohesion, or the Solar System bound by universal gravitation) can come to be and to be stable. In Newton's case, recognizing this distinction was part of recognizing the inadequacy of the mechanists' program for explanation; by ruling out any kind of interaction other than impact, the mechanists had ruled out just that method of investigation that the laws of motion had uniquely made possible, and that had made the existence and the nature of active forces a tractable empirical question. Moreover, as Kant had seen even more clearly than Newton, it was by making this distinction that the laws of motion had imposed a conception of causality on the phenomenal world, that is, a set of rules determining when and how the behavior of bodies requires a causal explanation. In short, what is definitive of special and general relativity as principle theories is equally definitive of Newton's theory, and definitive of the role that space-time structure plays in physical theories like Newton's and Einstein's. Relativity is no less a space-time theory than Newtonian space-time, and Newtonian space-time is no less a "theory of theories" than relativity. That is simply a reflection of the peculiar status of space and time in the world of classical – non-quantum – physics.

4.4 THE PHILOSOPHICAL MOTIVATIONS FOR GENERAL RELATIVITY

The transition from special to general relativity, as has been noted, had a number of philosophical motivations that no longer seem compelling. The philosophical defense of the relativity of motion, and of the "epistemological defects" of Newtonian mechanics and special relativity – any theory that restricts the relativity of motion to a special class of reference frames – involved some confusions about epistemology and, as we saw in Chapter 2, about what the earlier theories were really saying about space and time and their roles in physics. If we consider general relativity from a purely methodological perspective, as a theory to be judged by its consequences rather than by its motivations, none of this matters particularly; the theory is merely a hypothesis about space-time, and of no special philosophical interest except insofar as its success might support some metaphysical hypothesis. But if we look more carefully at its philosophical origins, we can see that, along with the somewhat misguided epistemological motivations, there really was a philosophical foundation, a compelling and fruitful conceptual analysis of the sort that gave rise to special relativity. By seeing this we will see that, after all, there is a particular philosophical significance to the argument for general relativity, though it does not lie precisely where the logical positivists, or even Einstein, had looked for it. It was not the general epistemological critique of distinctions between states of motion, but the dialectical engagement with the specific way in which Newtonian physics had made those distinctions, that yielded the basis for a new theory of space-time.

The first step to seeing this is to revisit the notion of a "general relativity of motion," and to see what this philosophical standpoint means, or fails to mean, for the construction of a theory of physical geometry. The demand for general relativity, or "the extension of the relativity principle to arbitrary coordinate systems," was assimilated by philosophers more or less as Einstein had first put it forward. The very meaning of the concept of motion, if there is one, must rest on change of relations to observable bodies. It would appear to follow that any theory that purports to distinguish different states of motion otherwise than relatively has, therefore, an inherent "epistemological defect." But until general relativity, the only empirically successful physical theories were defective in just this way. Newtonian mechanics did distinguish absolute rotation and acceleration, and so postulated a privileged class of reference frames in which rotation, along with force, mass, and acceleration, are invariant quantities. Special relativity offered a new conception of what is invariant – the speed of light – while "relativizing" simultaneity, length, and time, but in doing so it inevitably maintained the notion of a privileged class of reference frames and therefore preserved the epistemological defect. As we saw, Newton's contemporaries Huygens and Leibniz were credited with recognizing the defect, but had no serious proposals to overcome it; Mach, in the late nineteenth century, envisaged a genuine physical theory that would embrace the relativity of motion, but he was not able to contribute more than a visionary suggestion. It was only Einstein who found a way to realize the epistemological idea in physics. "Mach's principle" was originally a mere speculation that inertial effects might arise not from a body's absolute state of motion in space, but from its relation to the rest of the masses in the universe; Einstein had given physical content to this speculation, in the principle that the distribution of masses determines the local inertial behavior of moving particles.

Both of these theoretical steps were supposed to arise from the equivalence principle. It is not too much to say, indeed, that, independently of the equivalence principle, the idea of "generalizing" relativity of motion had no particular physical content. From the equivalence of gravitational and inertial mass, Einstein alone discerned that the distinct status of the inertial frame was fatally compromised. Galileo had proposed that all bodies fall with the same acceleration in the same gravitational field, and Newton had verified this principle to great accuracy; as Einstein knew, Eötvös had verified this with even greater accuracy as recently as 1895. But it follows that, in a local frame of reference, gravity will be impossible to distinguish from inertia. Suppose that the frame is a box at rest in a gravitational field where the characteristic acceleration is g; the downward accelerations of bodies – their weights toward the floor – will be the same as their inertial resistance to acceleration, if the box were accelerated upwards at the rate $-g$. Or, if the box is freely falling in a (sufficiently uniform) gravitational field, then every body in it will fall at the same rate, and to the observer falling with them, they will appear to have no acceleration at all; the frame under the influence of gravity will be indistinguishable from one that is acted on by no forces at all. For contrast, consider the behavior of magnetic bodies in an analogous situation: the observer could determine whether the frame was at rest in the magnetic field or accelerating upwards, by observing the behavior of a non-magnetic body such as a piece of wood. Since its inertia alone would determine its behavior, it would follow an inertial trajectory that could be distinguished from trajectories determined by the magnetic field. Only gravity has the characteristic that it cannot be distinguished from inertia. As Einstein later put it,

If there were to exist just one single thing that falls in the gravitational field differently from all the other things, then with its help the observer could recognize that he is in a gravitational field and is falling in it. If such a thing does not exist, however – as experience has shown with great precision – then the observer lacks any objective ground on which to consider himself as falling in a gravitational field.

Rather, he has the right to consider his state as one of rest and his surroundings as field-free with respect to gravitation . . . The experimental fact that the acceleration of fall is independent of the material is therefore a powerful argument that the relativity postulate has to be extended to coordinate systems which, relative to each other, are in non-uniform motion. (Einstein, 1920, p. 233)

From the collapse of the classical distinction between inertial and non-inertial frames, it seemed to follow immediately that the Lorentz invariance of special relativity must be replaced by a wider invariance group. The laws of physics must take the same form, not in a privileged class of coordinate systems, but in all possible coordinate systems; they must be generally covariant.

These philosophical motivations for a new physics are, evidently, destructive. By themselves, that is, they offer reasons to reject distinctions made by earlier theories, without suggesting any basis on which a new theory might be built, for example, a principle on which to construct the theory of non-Euclidean space-time geometry that we take general relativity to be. In the case of the special theory, the relativity principle was at the same time an invariance principle that constituted the metrical foundation for the Minkowski geometry. But these arguments for a general theory, even if they are persuasive in themselves, do not define a geometrical structure in the same straightforward way. The crucial link between the relativity of motion and non-Euclidean geometry was supposed to come from Einstein's thought experiment involving a rotating disc. If a disc rotates, special relativity implies that a measuring stick that rotates with it must be contracted in the dimension parallel to its velocity; since the velocity is greater as one approaches the edge of the disc, the increasing contraction of the stick will cause the lengths it measures to seem greater. As a result, to the observer at rest relative to the disc, the Euclidean relation between the diameter of the disc and its circumference will be disturbed; the co-moving observer must think that the geometry of the disc is non-Euclidean. That is, there will be an arbitrary choice to be made about the state of motion of the disc: either it is rotating and the stick is contracting, or it is at rest and has a non-Euclidean geometry. As Friedman (2002a) points out, this example enabled Einstein to see the possibility of representing physical fields through non-Euclidean geometry.[4] But it says nothing specific about how Euclidean geometry might represent the gravitational field.

Taken together, then, Einstein's general philosophical arguments, both destructive and constructive, fall short of motivating the general theory of relativity as we know it. We have already seen why the epistemological arguments against absolute rotation are misguided, even if it were true that general relativity had done away with any notion of a privileged state of motion; in fact, the theory preserves the idea of privileged state, in the special status accorded to gravitational free-fall. In effect, this fact only makes the prospect of "generalizing relativity" seem even more remote. Moreover, as Kretschmann first pointed out (1917) and Einstein came to realize soon after, the requirement of general covariance does not really amount to a destructive criticism of earlier theories. Instead, it can be satisfied by any theory that can be expressed in a coordinate-independent form, and this includes Newtonian space-time theory and special relativity; though the ultimate philosophical significance of general covariance is still a matter of controversy,[5] it became clear early on that, at least in the form that Einstein gave it, it had no inherent physical content. This is doubtless why Einstein retreated from the position that general covariance requires general relativity – and the general relativity of motion – to a somewhat weaker principle: that, among all generally covariant theories, we ought to prefer one that is most simply expressed in a generally covariant version. "The eminent heuristic significance of the general principle of relativity," he eventually concluded, "lies in the fact that it leads to the search for those systems of equations which are *in their generally covariant* formulation the *simplest ones possible*" (Einstein, 1949, p. 69). Special relativity and Newtonian mechanics may be generally covariant, but they can also be expressed equally (or more) simply by means of equivalence classes of privileged coordinate systems; general relativity has no such representation, and therefore is in this very narrow sense "singled out" by the requirement of general covariance. Yet this, too, is by itself a dubious reason to prefer general relativity. The other theories have privileged classes of coordinate systems because they have non-trivial global symmetries – that is, because they take space-time to be flat. General relativity, in contrast, has no symmetry group because it takes space-time to be non-uniformly curved. So, instead of a group of isometries, a general-relativistic space-time has only the automorphisms of the manifold itself, the functions that preserve the differentiable structure of the manifold. The a-priori philosophical preference for the theory that is simplest in its generally covariant form is, in effect, a philosophical prejudice against flat space-time. Therefore, in its philosophical spirit, such a preference seems to go against the entire tradition in physical geometry by which space-time curvature came to be understood as an empirical matter. From an empiricist perspective on physical geometry, if we prefer theories that are "only" generally covariant over those that have more restrictive invariances, it ought to be because we have empirical grounds for doubting that there are any global symmetries.

In sum, then, Einstein's general epistemological arguments for a general relativity of motion do little to help us understand the general theory of relativity, the theory that gravity is an expression of space-time curvature. What we need to understand is how Einstein derived a constructive principle for the new theory from his reflections on the equivalence principle. The epistemological arguments, as we saw, emphasized the need for a radical philosophical departure from earlier theories. But the discussions of the equivalence principle begin from the Newtonian theory of gravity, and the Newtonian method of distinguishing gravity from inertia. They uncover the empirical knowledge of the gravitational field that the theory is implicitly based on, and they separate the theory's implicit constructive principle – the principle underlying the construction of a Newtonian inertial frame – from the larger theoretical context in which Newton's theory had placed it. In this respect the argument for general relativity resembles the argument for special relativity, which had separated the empirical application of the concept of simultaneity from the theoretical context of Newtonian mechanics. In the case of general relativity, too, Einstein finds the basis for a new theory precisely in the old theory's practice of spatio-temporal measurement. Contrary to what Kuhn's analysis would lead one to expect, the argument for the new paradigm begins from inside the old.

To understand this point, we only need to reconsider the Newtonian procedure for determining the motions of the Solar System. It starts from a kinematical description of the motions, that is, a description of the accelerations of all the planets relative to the fixed stars. No assumption need be made about which planets are truly in motion. Neither must we assume anything about the motions of the stars, as long as they may be taken to be at rest relative to each other. That suffices to allow the determination of orbital parameters in a relatively theory-independent setting. Then, with this description in hand, we can introduce the dynamical theory in order to extract information about the forces at work within the system. By applying the laws of motion, their corollaries, and all the propositions proved from these in Book I of *Principia*, we can reason from the accelerations to the forces needed to produce them. In particular, we can reason from the characteristics of the orbits to the centers of those orbits, and from there to the forces needed to produce those orbits. This crucially involves the law of action and reaction, for otherwise it would be impossible to break down the total acceleration of any planet into the components contributed by particular other planets; the Earth's acceleration, for example, is the sum of its accelerations toward all the other planets, and each individual component of the total acceleration is part of an action–reaction pair involving some other planet. Only this assumption allows us to determine whether any further acceleration is at work, from some force whose source is unobservable or outside the system.

When we understand the mutual interactions among the planets, we are in a position to estimate their relative masses. In Newton's case, this was necessarily restricted to the planets with satellites, because only in those cases could he compare the accelerations they determine at given distances and so deduce the differences in mass. By this reasoning he estimated the ratios of the Sun's mass to those of Jupiter (1067 to 1), Saturn (3021 to 1), and the Earth, and was able to calculate that the center of mass of the entire Solar System would never be more than one solar diameter from the center of the Sun. Having found the center of mass, we have in principle determined an inertial frame: by Corollary IV to the laws of motion, the center of mass will be at rest or moving uniformly in a straight line. That is, the mutual actions of the bodies in the system will not change the state of motion of the center of mass. And having determined an inertial frame, we are in a position to say that the accelerations relative to the center of mass frame are the true accelerations.

One might suspect that there is some arbitrariness in this procedure, because of the hypothesis that the stars are a suitable frame of reference. But that is not really a hypothesis in the important sense – that is, it is not something that the procedure must simply take for granted as its basis. Rather, it is at most a kind of working hypothesis, for the dynamical reasoning can always lead to its rejection: if we don't succeed in resolving accelerations relative to the stars to their dynamical components, then we must infer that the stars are not a suitable frame after all. If we choose the Earth to be at rest, for example, we find that such a dynamical analysis does not succeed, for we discover accelerations (Coriolis and centrifugal effects) that do not belong to action–reaction pairs, and must be attributed to the state of motion of the reference frame itself. What seems like a basic presupposition for such a process is, instead, a mere starting point that itself undergoes a critical analysis as the investigation proceeds; this is another aspect of the situation that really does deserve to be called dialectical. In short, whether the fixed stars are a suitable frame is considered an empirical question in this analysis, answered by the success of the analysis itself. This situation is not affected by the criticisms of Mach; Mach himself acknowledged that such a process of analysis had taken place, and had led to the enlargement of the frame of reference for physics from a small, nearly flat region, to the Earth itself, and finally to the fixed stars in the hands of Galileo and Newton (Mach, 1889, p. 215). For Mach, however, once Newton had discovered the interdependence of interactions and accelerations relative to the fixed stars, the analysis could go no further. That accelerations relative to the fixed stars follow Newton's rules is an empirical claim, and any extrapolation beyond it would be empty metaphysics. For Newton, in sharp contrast, this was just a step in the application of a universal law, and therefore the beginning of a process that must be completely open-ended.

The real arbitrariness in this procedure stems not from the initial choice of a reference frame, but from the peculiar nature of gravitation. For the dynamical reasoning just outlined involves separating the purely inertial component of every accelerated motion from components of acceleration due to gravity. If we could not do this, we would be in no position to say that the center of mass of a system of bodies moves uniformly, or that accelerations relative to that center are true accelerations. Yet this is precisely what the equivalence principle prevents us from doing. Because gravity acts equally on all bodies, a system of bodies that is falling in a gravitational field will be indistinguishable from one that is moving uniformly, and therefore the accelerations relative to the center of mass, as determined by the Newtonian analysis, cannot be known to be the true accelerations. We have already seen, of course, that Newton was aware of this situation, having inferred it himself from Corollary VI. Having exploited it for the analysis of Jupiter and its moons, to show that the analysis does not depend on whether the center of mass of that system is in inertial or freely falling motion, he could hardly have missed the application to the entire Solar System that he remarked upon in *The System of the World* (see Chapter 2, earlier). (See Figure 6.) He could, however, treat it as a merely practical limitation. If the true accelerations of Jupiter and its moons can only be studied in the context of the entire Solar System, then the true acceleration of the Solar System itself can only be studied within a still more encompassing system, and so on. This series can only come to an end when we can be confident that all matter in the Universe has been included in one dynamical system; in that case, the law of action and reaction will eliminate the possibility of a further acceleration of the system's center of mass, for by hypothesis there will be nothing with which the system might be interacting. As Kant pointed out, this implies that Newton's own notion of determining "the true motions," by a dynamical analysis of the forces at work, represents an ideal limit of scientific inquiry (see above and Chapter 3, earlier).

Newton's idea of determining true motion, then, like that of absolute simultaneity, has to be regarded not as an empirical procedure, but as an abstraction from the empirical procedure. More precisely, the notion

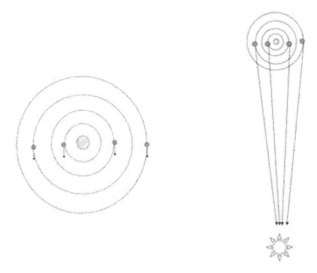

Figure 6. Newton's Corollary VI: the system of Jupiter and its moons can be treated as an inertial system, since all the gravitational accelerations toward the Sun are (given the immense distance from the Sun) very nearly equal and parallel, though in fact they gradually converge toward the center of the Sun.

of determining the true acceleration abstracts the empirical procedure of determining acceleration relative to the local center of mass – as Newton accomplished for the Solar System and its sub-systems – from any empirical circumstances in which it might practically be carried out. It makes implicit use of what became the equivalence principle, by exploiting the universality of free-fall together with Corollary VI, but it does not, as Einstein often noted, incorporate it into the structure of the theory. It is simply a matter of fact that, among the forces of nature, there is one that behaves like the sort of force described in Corollary VI, affecting all bodies in such a way that its very existence can be practically ignored for certain kinds of dynamical problem. It is also simply a matter of fact that it is only this force that allows us to study the interactions of the planets, to acquire any estimate of their masses, and so to have any hope of determining a dynamical frame of reference – of solving "the frame of the system of the world." We need not infer from this, as Kant did, that the existence of the gravitational force is a synthetic a-priori principle. But we might infer, as Einstein did, that the tasks of identifying inertial motion and measuring the local gravitational field are intimately bound together in a way that the Newtonian framework does not quite express.

Indeed, it was probably impossible to express this connection between inertia and gravitation until it could be represented in the framework of space-time. In that context, the Newtonian program is to comprehend every space-time trajectory as a deviation from a space-time geodesic; the center of mass of any isolated dynamical system must follow a space-time geodesic, and the curved trajectories of its parts must be comprehended as accelerations relative to the center of mass, and explained by their mutual interactions. By the equivalence principle, however, a free-fall trajectory will be indistinguishable from a geodesic. Therefore the acceleration of a falling body relative to the local center of mass may be, for all we are able to determine, merely a relative acceleration of two free-fall trajectories. The contradiction that Einstein identified – that two frames that may be regarded as inertial nonetheless have relative accelerations – cannot be resolved except in the setting of space-time geometry; only through the identification of the gravitational field with the space-time metric can we represent the objective structure underlying the perspectives of these frames. (See Figure 7.) This is a significant difference from the case in special relativity, where the "three plus one"-dimensional representation – Einstein's 1905 account, in short – adequately represents the theory, even if Minkowski space-time puts it in a much clearer perspective. This is because all inertial frames are related by the Lorentz symmetry group, so that the symmetry group itself expresses the structure of space-time, whether we choose to see it in those terms or not. Where the inertial frames are relatively accelerated, and their acceleration depends on the distribution of mass, there is no symmetry group, and so no way of simply representing the theory by its "relativity principle." It is plausible to say that special relativity is "about" the equivalence of a certain class of frames, instead of speaking about the structure of space-time, because the equivalence of those frames defines the structure of the theory, and simultaneously defines a structure for space-time. But knowing that freely falling frames are equivalent to one another tells us, at most, the local structure of space-time; to the extent that special relativity is locally approximately true in those frames, we can say that the space-time is locally Minkowskian. To know something of the larger-scale structure of space-time, and its dependence on the distribution of matter, we need to know how to interpret the relative accelerations of the local inertial frames, in such a way that they serve as measurements of the space-time curvature.

It is helpful to see this problem from what might be called, very loosely and somewhat facetiously, a Hegelian perspective. The equivalence principle obliges us to confront what appears to be, within the framework of flat

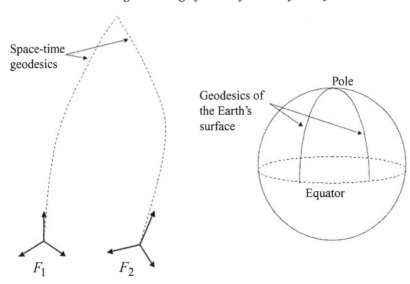

Figure 7. Free-fall as an indicator of space-time curvature: the local inertial frames F_1 and F_2 are in general not inertial relative to one another, since they may converge or diverge. This relative acceleration is a measure of space-time curvature, just as the convergence of geodesics (e.g. lines of longitude) on the Earth exhibits the curvature of the Earth's surface.

space-time, a contradictory situation: different local inertial frames separately satisfy empirical criteria for being inertial frames, yet are non-inertial relative to one another. In some sense this circumstance must undoubtedly undermine the distinct status of the classical inertial frame, but it is not immediately obvious how and why. Einstein thought that this circumstance supported the "extension of the relativity principle," in part because of his way of thinking about coordinate systems in general; if the class of privileged frames cannot be distinguished from falling frames, then the restricted relativity principle has to be extended to include all possible coordinate systems in all possible states of motion. But to do so is to set aside the lesson of the equivalence principle, that is, the very peculiar character of free-fall and its peculiar relation to inertial motion. The "general relativity of motion" that results is then something like what Hegel referred to as "the night in which all cows are black," the reconciliation of differences simply by blurring them. In a genuine dialectical resolution, however, those very differences would be revealed as telling us something about the more comprehensive reality, the larger view of which they represent limited local perspectives. There is a direct analogy between this case and that of local perspectives on the surface of the Earth. On the one hand, there is everywhere a privileged vertical direction, a direction that locally satisfies empirical criteria for being vertical. On the other hand, however, it will be observed that the directions thus identified at different places, e.g. by the directions of particular stars at particular times, are not vertical relative to one another. It might seem philosophically reasonable to adopt therefore a kind of relativity principle regarding direction. Yet there is some insight to be gained from the contradiction, if we can only see that there is a larger perspective in which each of these locally privileged directions has a natural place, and its locally privileged character makes sense as a limited perspective on something larger. What needs to be recognized, in short, is the spherical shape of the Earth and the convergence of all vertical lines on its center. In the same way, the inertial character of reference frames that are non-inertial relative to each other, as implied by the equivalence principle, must be seen as revealing something about the larger-scale structure on which they provide seemingly conflicting perspectives. This can only come about if instead of thinking about inertial frames and what they tell us about the relativity of motion, we think about inertial trajectories and what their divergences tell us about the curvature of space-time. Contrary to the spirit of the "general relativity of motion," the general theory of relativity itself requires us to acknowledge the uniqueness of free-fall trajectories, and to understand what it reveals about the larger structure of space-time.

4.5 THE CONSTRUCTION OF CURVED SPACE-TIME

This last question appears to receive little emphasis in Einstein's fundamental paper, in comparison with the destructive epistemological arguments that we have considered. But it is, in fact, the crucial point in Einstein's reasoning: its answer determines the application of the "generally covariant formalism," the geometry of curved space-time, to the physics of gravitation. On its answer, indeed, both the physical significance of general covariance and the geometrical significance of the equivalence principle entirely rest. What needs to be established is that the physical uniqueness of free-fall corresponds to a unique element in the geometrical formalism. From the retrospective of the logical positivists, again, this would require the setting down of an arbitrary stipulation, coordinating geometry and physics or, specifically, coordinating the geodesic of space-time with the trajectory of a falling body. But from Einstein's perspective, the link between the two is not presented as a stipulation at all. Rather, it is presented as a kind of discovery, at once physical and mathematical, that what is distinct about free-fall corresponds to what is distinct about geodesic trajectories: the only objectively distinguishable state of motion corresponds to the only geometrically distinctive path in a generally covariant geometry (Einstein, 1916, pp. 29–30; 41–2). In other words, free-fall motion satisfies objective empirical criteria for a privileged state of motion, just as the geodesic satisfies an objective mathematical criterion that does not depend on the choice of coordinates. We can choose a coordinate system in which a given free-fall motion is not rectilinear and uniform, and therefore introduce a gravitational field as the cause of its (relatively) non-geodesic motion. But this only shows that the Newtonian gravitational field is essentially arbitrary. As we saw, in the Newtonian procedure, no *actual* measurement of gravitational acceleration can ever really measure the deviation from a geodesic in flat space-time – that is, the measurement is never the absolute acceleration in Newton's sense, but the relative acceleration of trajectories that may both be regarded as freely falling. Thus the objective empirical facts are the characteristics of the free-fall trajectories themselves, and the objective (covariant) geometrical fact is the mutual divergence of their local coordinate systems.

It should be clear from the foregoing, then, that general covariance expresses an interesting physical fact only in conjunction with the equivalence principle. In order to identify a Newtonian inertial frame, and to identify the gravitational accelerations of bodies with respect to that frame, we have to determine its center of mass. But since its center of mass, again, may be itself in gravitational free-fall, the accelerations that we measure are in fact only relative accelerations, that is, accelerations of certain freely falling bodies relative to some other free-fall trajectory. Therefore the determination of a Newtonian inertial frame is essentially an arbitrary choice of coordinates – an arbitrary decision that a certain free-fall trajectory is really a Newtonian inertial motion, and an arbitrary decision that its particular coordinate system really is an inertial frame. By a similar analysis, we can see the arbitrariness in the Newtonian field equation, that is, the Poisson equation relating the gravitational potential to the mass density. For the gravitational potential itself is empirically measurable only as relative acceleration – the tidal accelerations of a given system of falling bodies. Since these accelerations will be independent of the state of motion of the entire system, the magnitude of the gravitational potential is always a matter of arbitrary choice; its value evidently depends on our initial stipulation of a coordinate system. Even if the Newtonian theory can be expressed in a generally covariant form, the equivalence principle implies that its most fundamental theoretical quantities are – not in virtue of the theory's mathematical form, but in empirical fact – coordinate-dependent. So, taken together with the equivalence principle, the demand for general covariance does have some direct physical significance after all. To grasp its significance, we first had to grasp that it arises, not from a general epistemological fact about spatio-temporal measurement, but from physical facts about the gravitational field and the ways in which we are capable of measuring it.

The decisive part of Einstein's argument, then, is not the epistemological critique directed against the Newtonian idea of motion. Rather, it is the conceptual analysis directed against the Newtonian distinction between inertial and gravitational motion, and against the Newtonian procedure for determining the action of gravitational forces. The analysis shows that the Newtonian approach does contain an implicit distinction between the objective and the arbitrary quantities, but one that is not really captured by the physical principles by which Newton defines the concept of a gravitational field. For that definition rests on the distinction between inertial and forced motion that is implicit in the laws of motion: gravity causes a deviation from the privileged trajectory, or geodesic of flat space-time. But, as Einstein's arguments show, that is not the conception of a privileged trajectory that is *actually in use* in the analysis of any real system; the one that Newton exploits is just the one that Einstein derives from the equivalence principle, that is, the trajectory of the center of mass of an isolated system that may as well be freely falling. So the relative acceleration that is measured is never that of a falling trajectory relative to a uniform rectilinear trajectory, but that of one falling trajectory relative to another. It would hardly make sense, then, to say that Einstein is arbitrarily coordinating a type of observable motion with the geometrical notion of a geodesic. Rather, he has found the conception of geodesic that is implicit in our actual knowledge and practice.

Characterizing Einstein's conceptual analysis in this way might seem to bring it closer to Mach's analysis of inertia. For, according to Mach, our actual practice of measuring the interactions among the planets always depended implicitly on their accelerations relative to the fixed stars, and so a genuine analysis of the concept of inertia must be one that reduces its meaning to this empirical and practical basis (see Chapter 2, earlier). Newton's understanding of the dynamical analysis, as we observed, involved an abstraction from every empirical case in which it might be carried out, so that absolutely any material circumstances – including a body rotating in an otherwise empty universe – could be seen as a special case of the universal laws that define inertia and force. Properly understood, Mach's fundamental objection was just to this abstract view: our actual knowledge, he insisted, was limited to the link between interactions and motions relative to the fixed stars. While Mach did not really have a theory of the origins of inertia, he saw the mere possibility of such a theory as a good enough reason to reject an abstraction like Newton's; the seeming compatibility of all our evidence with such a theory just as well as with Newton's offered a further reason. But Einstein's approach is not really Machian in this sense. For his approach, too, involves an abstraction from the empirical characteristics of motion relative to the fixed stars. The crucial distinction is that, where Newton considered what the system we are acquainted with has in common with an ideal, truly isolated system, Einstein's abstraction involves considering what this system has in common with every local system that may be treated as freely falling. His application of the equivalence principle thus identifies the universal distinguishing feature of freely falling local systems: that they satisfy all the empirical criteria for being inertial systems, without being integrable into any single global inertial system. Precisely this non-integrability enables the comparison of inertial frames to reveal something about the structure of the gravitational field. Where Newton's abstraction had the function of separating gravity from inertia, Einstein's abstraction reveals their underlying unity.

From this analysis, the step to space-time curvature is remarkably simple. The relative behavior of geodesics is, simply, a defining characteristic of the curvature of a space: in a flat space, geodesics do not exhibit relative acceleration ("geodesic deviation"), whereas in a curved space, the relative acceleration of geodesic trajectories provides a measure of the curvature, as expressed mathematically in Riemann's curvature tensor. The identification of curvature with gravity follows more or less automatically: the geodesic deviation is measured by the relative acceleration of falling bodies, which, as we just saw, is our empirical measure of the gravitational field. The search for field equations, relating the space-time metric to the distribution of matter and energy, was historically a difficult and involved one,[6] but the fundamental idea is now almost obvious. For once we re-assess the Newtonian field equation in light of the unity of inertia and gravity, we know that the empirical basis of the equation is just the relation between mass distribution and the relative accelerations of falling bodies; and once we learn to interpret the latter as geodesic deviation, we understand that the real content of the field equation must be the relation between mass distribution and space-time curvature. As Einstein put it, once we know that local inertial frames are relatively accelerated, so that "we are no longer able by a suitable coordinate-choice to make the special theory of relativity valid in a finite region, we will have to hold fast to the conception that the $g_{\mu\nu}$ describe the gravitational field" (1916, p. 17). What had seemed to be a contradiction, then, is now resolved: the apparently incompatible viewpoints of different inertial coordinate systems are just local perspectives on space-time that is curved on a larger scale. The divergence between the coordinate systems is, instead of an irreconcilable conflict, a precise quantitative measure of the space-time curvature.

From the foregoing we can see that the essential point of general relativity, the identity of gravitation with space-time curvature, is entirely independent of special relativity. And this is why Newton's theory of gravitation, within a space-time framework defined by absolute simultaneity and Euclidean spatial geometry, can be formulated as a theory in which falling bodies follow the geodesics of a curved affine structure (see Trautman, 1965; Malament, 1986; Stachel, 2002c). Therefore, to call Einstein's argument a conceptual analysis might seem to imply that he was revealing something that was there all along in Newton's theory, and that only the lack of mathematical formalism prevented Newtonians from seeing. But this would be to ignore the contingency of conceptual analysis on the evolution of physical theory and observation. The indistinguishability of inertial motion and free-fall does not *necessarily* undermine the global determination of space-time geometry; it does so only on the assumption that there are no physical phenomena *independent* of gravitation that might serve to measure the background space-time geometry. It is not self-evident in Newtonian physics, for example, that light rays must be subject to gravitational forces, or, therefore, that electromagnetic phenomena must be subject to Corollary VI just as mechanical phenomena are. In principle, electromagnetic or other phenomena might exhibit a background geometrical structure that is distinguishable from the gravitational field; we might then be able to measure the acceleration of a freely falling particle relative to some other trajectory that is not dependent on gravity – just as we can, in fact, measure the acceleration of a particle in a magnetic field relative to the inertial trajectory of a body that is not affected by magnetism. For this reason, it was a crucial step in Einstein's reasoning to extend the equivalence principle to all physical interactions: "But this view of ours [i.e. of the equivalence of a system **K** at rest in a homogeneous gravitational field, and a system **K'** that is uniformly accelerating] will not have any deeper significance unless the systems **K** and **K'** are equivalent with respect to all physical processes, that is, unless the laws of nature with respect to **K** are in entire agreement with those with respect to **K'**" (Einstein, 1911, p. 101). The confirmation of general relativity by the deflection of starlight, in the celebrated eclipse observations of 1919, was so important precisely because it argued that light is, after all, subject to the equivalence principle: that is, the paths of light rays cannot be used to separate the gravitational field from some flat background geometry. Therefore this argument had, just at that stage in the development of physics, a certain force that it could not have had much earlier. So Einstein's conceptual analysis is not analysis in Kant's sense, that is, analysis of what is contained in a fixed concept; nor is it the mere reduction of the concept to its supposed observational consequences. It is, instead, an analysis of the *evolving* role that the concept plays in an *evolving* body of theory and practice.

Thus general relativity does not, after all, reduce the concept of motion to purely observable relations, or blur the distinctions among different states of motion. Instead, it begins from the local inertial system as a privileged frame of reference, and the seeming paradox that different inertial systems may be relatively accelerated. But the theory unites the perspectives of different inertial systems in a more comprehensive framework, in which their mutual differences reveal the structure of space-time on a larger scale. Einstein's argument for the theory, therefore, follows the dialectical pattern that we noted in his argument for special relativity. Like the passage from Newtonian space-time to special relativity, the passage from special to general relativity begins with a measurement procedure that makes perfect sense according to the old theory, but whose application in frames of reference in certain states of relative motion leads to surprising results. The procedure for identifying the Newtonian gravitational field is essentially carried over unchanged into the new theory; the radical change comes from the recognition that it can be carried out even in frames of reference that are relatively accelerated – i.e. that the criteria for an inertial frame can be satisfied even in frames that are accelerated relative to one another. As in 1905, there is an apparent contradiction here whose resolution requires us to reintepret the basic concepts of the theory.

In at least one sense, however, the resolution to which it leads is less radical than that of special relativity. The passage to special relativity reveals that what had been thought to be absolute has turned out to be relative; the precise meaning of this is that the symmetry group of space-time has changed, so that certain quantities that were invariant in the old theory are frame-dependent in the new, and vice versa. The "relativization" that occurs in the passage to general relativity cannot be described in such precise terms. Because local inertial frames are likely to be relatively accelerated, they don't stand in the same simple relation to one another that characterizes inertial frames in Newtonian and special-relativistic space-time; the shift from one to another cannot be a rigid displacement of space-time, as it would be if the frames were determined by a symmetry group. This is the true content of Einstein's remarks about the need to "free oneself of the belief that coordinates must have an immediate metrical significance" (1949, p. 66); we have to free ourselves of the a-priori conviction that coordinate systems directly express the possibility of rigid spatial and temporal displacements, and that coordinate transformations correspond to such displacements in the simple and direct way that they did in flat space-time. In short, inertial frames in general relativity have the same structure, locally, as in special relativity; what has changed is their relation to one another. Special relativity, in relation to Newtonian mechanics, meant a change in the structure of the inertial frames: something that was supposed to be true in all inertial frames (e.g. that certain events are simultaneous) turned out to be dependent on the choice of frame, whereas all frames would now agree on something (the velocity of light) that could not possibly have been an invariant of the old structure. General relativity, in relation to special relativity, implies that what is true in an inertial frame is true only locally – which is to say that there really are no inertial frames, even if certain fairly small regions of space-time may be hard to distinguish from inertial frames. Instead of saying that what was absolute turns out to be relative, then, we should say that what was global turns out to be local. It is somewhat misleading, then, to infer from the equivalence principle that "the relativity postulate has to be extended to coordinate systems which, relative to each other, are in non-uniform motion." If this is in one sense an extension of the relativity principle, it is in another sense a restriction of it: it implies that the equivalence of local inertial frames – just because they are local – can no longer be seen as defining a symmetry of space-time.

4.6 GENERAL RELATIVITY AND "WORLD-STRUCTURE"

Emphasizing the conceptual analysis occasioned by the equivalence principle, as the essential part of Einstein's reasoning, gives us some insight into the emergence of general relativity as a constructive theory of space-time geometry, rather than as a somewhat questionable philosophical critique of older theories. Just for that reason, however, the foregoing account might seem unhistorical. It might appear merely to separate, from the confusing jumble of Einstein's philosophical motivations, the ideas that best conform to our present understanding of general relativity. But my aim has not been to isolate what now seems the most reasonable foundation for Einstein's theory. Rather, I have tried to understand how Einstein's jumble of motivations gave rise to an empirical theory, one that made the structure of space-time a matter of empirical investigation. Such an understanding is further vindicated by the fact that, in Einstein's own time, it was shared by those who understood the theory particularly clearly – that is, those who gave the theory the mathematical form that expressed its physical content most clearly, and that played the most influential role in its wider assimilation. As Minkowski had done with special relativity, Eddington and Weyl, within a few years of the advent of general relativity, presented both the mathematical formalism and the physical content of the theory, in the first perspicuous and comprehensive texts on the theory (Weyl, 1918; Eddington, 1920). These works noted the epistemological criticism of earlier theories that had been such an important part of Einstein's presentation, but they did not especially emphasize it, or identify it as the special philosophical import of general relativity. Instead, they presented general relativity in a manner that made its continuity with its predecessors much easier to see. In this way they revealed general relativity as a theory, not of relativity, but of "world-structure" – the dynamical structure that unified the physics of gravitation with the geometry of space and time.

This view has never been fully assimilated in the familiar philosophical debates concerning space and time. Or, more precisely, the debate has only considered the very broadest aspect of such a view, that it is a generally realistic interpretation of space-time structure – a view that, more than the positivists had realized, deserves to be taken seriously as a possible and even plausible view of the metaphysics of general relativity. But the emphasis on "world-structure" was more than a metaphysical position or hypothesis. It rested on an understanding of our knowledge of space and time, in particular the relation between physical geometry and our empirical assumptions about physics, that illuminated the structure and the significance of general relativity in a way that other philosophical interpretations did not. In doing so it brought Minkowski's approach to physical geometry into harmony with a dynamical view of space-time. To say this is not to insist upon a particular interpretation of general relativity.[7] It is merely to acknowledge that this approach to the theory answered certain questions about it – especially, how it stands in relation to earlier theories, and precisely what it asserts about the nature of space, time, and motion – that the more prominent interpretation tended to obscure.

Weyl's *Raum-Zeit-Materie* (1918) presented Newtonian physics, electrodynamics, special relativity, and general relativity within a general mathematical framework that, at the same time, placed their philosophical relations in a coherent perspective for the very first time. The progression from

Newton's theory to Einstein's, as Weyl presents it, is not primarily a movement from naïve absolutism to relativism, or the gradual erasure of every trace of "physical objectivity" from space-time, but a deepening understanding of world-geometry and its relations with the dynamical properties of matter. While he shared in the general enthusiasm for Einstein's "relativistic" philosophical viewpoint and seemed to take Mach's principle fairly seriously, he also emphasized that such a general epistemological viewpoint, by itself, offered little insight to the structure of the theory. For Weyl, the true significance of general relativity lay in "the assumption that the World-metric is not given a priori, but the quadratic groundform is to be determined by matter through generally invariant laws" (Weyl, 1918, pp. 180–1). The requirement of general covariance he regarded as "essentially mathematical" rather than as expressing the physical content of the theory; thus the "essential kernel," he thought, was to be found "less in the requirement of general invariance than in this principle [that gravitation is a mode of expression of the metric field]" (Weyl, 1918, p. 181). He could not, therefore, regard Mach's principle as Einstein (at first) and the logical positivists did,[8] as merely the physical application of an undeniable epistemological principle; it has to be seen as a physical hypothesis about the source of centrifugal effects and, moreover, one that is only partly compatible with the true content of the new theory. Thus he pointed out that the phenomena associated with "absolute rotation" are "in part an effect of the fixed stars, relative to which the rotation takes place" – adding in a footnote, "In part, because the mass-distribution in the world does not uniquely determine the metric field . . ." (Weyl, 1918, pp. 175–6). In other words, the metrical field has a certain independence of the distribution of matter, and therefore the states of motion of bodies with respect to the field have a corresponding independence of the distribution of matter; general relativity thus does not fulfill the Machian aim of reducing all motion to the relative motion of observable bodies.

Later, in his survey article, *Philosophie der Mathematik und der Naturwissenschaften* (Weyl, 1927), he went much further in criticizing the Machian emphasis on the relativity of motion, even stating explicitly that it is a hindrance to the understanding of Einstein's theory:

Incidentally, according to the general relativity-postulate, without any basis in a world structure, the concept of relative motion of several bodies is left hanging in the air just as much as the concept of absolute motion of a single body . . . Thus a solution of the problem consistent with the tendency of Huyghens and Mach, which seeks to eliminate the world-structure, is impossible. (Weyl, 1927, p. 74)

This passage makes a point that the relationalist tradition, from Newton's time through the nineteenth century, had largely ignored: that the relative motions are not purely phenomenal, and therefore are not epistemologically privileged over geometrical structure. On the contrary, the description of spatial relations, and their changes over time, presupposes a degree of geometrical structure just as the description of absolute motion does. Moreover, from the standpoint of special relativity, the presuppositions underlying relative motion seem questionable and naïve. Only on the assumption of absolute simultaneity can we even identify the situation of bodies, in the sense of Leibniz or Mach, in order to measure how it changes over time. Thus some conception of world-structure underlies *any* attempt to understand motion, even those that claim to reduce motion to observable relations.

The philosophical problem for physics is therefore not to eliminate the world-structure by reducing it to some epistemologically sound foundation – as if it were nothing but an empty metaphysical abstraction from the objective empirical relations. The problem, rather, is to grasp the nature of the world-structure through the physical phenomena that reveal it. What had made the structure seem otherworldly, "metaphysical" in the pejorative sense, was the Newtonian assumption that such a structure could stand apart from all physical interactions, determining their course without being affected by them in any way. What was "unsatisfying" was that "something that has such powerful effects as inertia . . . is supposed to be only a rigid geometrical property of the world, fixed once and for all . . . Therefore the solution is given as soon as we are resolved to acknowledge *the inertial structure as something real, that not only exerts effects upon matter but also suffers such effects*" (Weyl, 1927, p. 74). The structure that reveals itself in the motions of falling bodies is, at the same time, in an interaction with bodies that reveals itself in the dynamics of the gravitational field. Where Einstein had seen general relativity as taking away "the last remainder of physical objectivity" from space and time (1916, p. 13), Weyl sees it as granting space-time geometry the same kind of reality that we grant to every other physical field.

Evidently, then, Weyl's philosophical perspective on general relativity was very similar to Minkowski's on special relativity: he was convinced that the philosophical emphasis on relativity as a general epistemological principle, and the emphasis on equivalence of coordinate systems as a physical principle, only obscured the underlying geometrical structure that is the theory's true content. Like Minkowski, he did much to translate the talk of coordinate systems and transformations into a coherent mathematical account of that underlying structure. But what interests us most for the present is his discussion of how we come to know the structure of space-time, or more precisely, how we come to understand certain characteristic phenomena as having some kind of geometrical significance. It is in this discussion that another kind of kinship with Minkowski emerges, for Weyl defends a very similar view of the direct relationship between geometry and its physical interpretation. The relationship begins from the most elementary ideas that we form about the events that we experience:

> One already attributes a definite *structure* to the four-dimensional extensive medium of the external world, if one believes in a division of the universe into an absolute space and an absolute time, in the sense that it is objectively meaningful to say of any two separate events, narrowly localized in space-time, that they are happening at the same place (at different times) or at the same time (at different places). (Weyl, 1927, p. 65)

When we begin to introduce quantitative considerations regarding length or duration, we implicitly introduce a more complicated structure: "One attributes to the world a metrical structure when one assumes that the equality of time-intervals and congruence of spatial figures have an objective meaning" (Weyl, 1927, p. 66). There is a natural interpretation of spatio-temporal structure, then, because assumptions about the structure are implicit in the way that we make certain fundamental empirical distinctions.

With the introduction of dynamical laws, we come to understand another kind of structure, the inertial structure. "The experiences which prove the dynamical inequivalence of different states of motion teach us that the world bears a structure" (Weyl, 1927, p. 70), namely the affine structure of space-time. According to Weyl, Newton had rightly recognized this connection, but "this inertial structure was not correctly interpreted by the concept of absolute space" (Weyl, 1927, p. 70), because where absolute space rests on the distinction between motion and rest, the meaningful distinction (see Chapter 2, earlier) is between uniform motion and acceleration. There are three points of interest in Weyl's remarks about Newton. First, in contrast to Einstein and the logical positivists, he does not object to Newton's belief that dynamics can tell us something about the objective nature of space and time; he only criticizes the concept of absolute space as the wrong way of characterizing the structure. Second, as we noted above, his only general philosophical objection to the reality of inertial structure, and to the claim that we can grasp it through inertial forces, is that the classical notion removed the structure from all interaction with physical fields. Third, these remarks make even more explicit how Weyl regards the problem of interpreting geometrical structure: we do not choose an interpretation for an abstract mathematical structure, but we try to interpret characteristic physical phenomena through a conception of geometrical structure – as, for example, absolute space was an (inadequate) attempt to interpret the evident dynamical distinction between uniform and accelerated motion. Weyl was convinced, as we noted, that there were a-priori philosophical reasons for preferring an inertial structure that is dynamical rather than absolutely fixed. But the true justification for accepting such a structure is that it interprets for us the motions of falling bodies: if these are indistinguishable from geodesic motions, then the dynamics of the gravitational field – its dependence on the distribution of matter – reveals to us the dynamics of the affine structure of space-time. Or, in other words, if freely falling frames behave like inertial frames, then inertial frames are purely local, and the dynamics of the gravitational field is expressed by the divergences among different inertial frames. Minkowski's and Newton's space-times – flat space-times that allow for a privileged class of *global* inertial frames – take no account of the dynamical relation that is revealed by free-fall, and so they attempt to impose on all of space-time a framework that can only be applied to the smallest regions. In order to understand the significance of the identity of inertia and gravitation, in sum, we need to see that it directs us to interpret free-fall trajectories as revealing the world-structure on a larger scale.

Weyl's understanding of general relativity, and, more broadly, of the connection between physical phenomena and geometrical structure, had a direct influence on Eddington (1918). That influence is particularly obvious from the philosophical discussions of the theory that he wrote for non-specialist readers, in *Space, Time, and Gravitation* (1920). For our purposes, however, what is most relevant and most revealing is the case he presented to physicists for general relativity as an empirical physical theory – as the kind of theory in which physicists could have the kind of confidence that they had in Newton's theory of gravity, the kind of theory whose connection with empirical evidence could be made evident, despite the unfamiliar and bizarre-sounding mathematical framework in which it was expressed. This case became best known, and most influential, through the text, *The Mathematical Theory of Relativity* (1923). But in 1918 – before the historic eclipse expedition to measure the deflection of light by the Sun – Eddington presented a "Report on the relativity theory of gravitation" to the Physical Society of London (Eddington, 1918). Here we can see especially clearly that, for Eddington, an inseparable part of the scientific case for the theory was a philosophical argument about our knowledge of space and time. If the interpretation of gravity as space-time curvature was to make any sense, Eddington apparently recognized, the question of space-time curvature itself would have to be made sense of as an empirical question; space-time geometry would have to be understood as a matter for objective measurement. In contrast to Einstein's view of space and time as lacking "physical objectivity," and of space-time coordinates as lacking "direct metrical significance," Eddington tried to articulate the sense in which coordinate systems and their application really do tell us something physically meaningful about the nature of space and time. This task required more than exhibiting a spectacular predictive success of the theory, such as the light-bending result; it required a compelling argument that the basic concepts of curved space-time geometry correspond to measurable physical magnitudes.

There are, to be sure, passages in which Eddington seems to embrace the epistemological view of Einstein, and in particular Einstein's view of what can be regarded as a measurable physical magnitude; he suggests that the equivalence principle furnishes the means of extending relativity in just the way that Einstein suggested, so that only local relations are observable, and a gravitational field may be "transformed away" by a suitable choice of coordinates.

It will be seen that this principle of equivalence is a natural generalization of the principle of relativity. An occupant of the projectile [in a gravitational field] can only observe the *relations* of the bodies inside to himself and to each other. The supposed absolute acceleration of the projectile is just as irrelevant to the phenomena as uniform translation is. The mathematical space-scaffolding of Galilean axes, from which we measure it, has no real significance. (Eddington, 1918, p. 20)

Eddington emphasizes, however, the "limitation of the Principle of Equivalence," namely that to transform away a gravitational field is only possible for an infinitesimal region of space-time. Of course, Einstein frequently made this point (e.g. 1916, p. 41; 1922, pp. 63–4). But Eddington gives more emphasis to what this limitation implies for our knowledge of space and time – more precisely, for our ability to determine the structure of space and time by empirical measurement.

. . . [For] an infinitesimal region the gravitational force and the force due to a transformation correspond; but we cannot find any transformation which will remove the gravitational field throughout a finite region. It is like trying to paste a flat piece of paper on a sphere, the paper can be applied at any point, but as you go away from the point you soon come to a misfit . . . The impossibility of transforming away a gravitational field is, of course, an experimental property; so that, in spite of the principle of equivalence, there is at least one means of making an experimental decision.

Space-time in which there is no gravitational field which cannot be transformed away is called *homaloidal* . . . Our space is not like that, though we believe that at great distances from all gravitating matter it tends toward this condition as a limit. (Eddington, 1918, p. 22)

Eddington is pointing out the fact, also emphasized by Einstein, that the local character of inertial frames makes for discrepancies between the inertial coordinates defined at different points at space-time; this is just another way of saying that local inertial frames can be in non-uniform motion relative to one another, and the coordinate system of one cannot be extended to embrace that of another, as one would expect to be able to do in flat space-time. For in flat space-time every local inertial frame is extendible into a global frame, with respect to which every other local inertial frame remains an inertial frame. But Eddington is also arguing that this discrepancy between local inertial frames, properly understood, reveals something objective about the underlying structure of space-time, namely its non-homaloidal character.

This way of thinking leads Eddington to quite a different view from Einstein's of the relation between general relativity and the Newtonian view of space and time – and the relation between general covariance and "general relativity" in the broader philosophical sense. The difference is expressed sharply in Eddington's discussion of absolute space, and why and to what extent it is rejected by general relativity:

Although we deny absolute space, in the sense that we regard all space-time frameworks in which we can locate natural phenomena as on the same footing, yet we admit that space – the whole group of possible spaces – may have some absolute properties. It may, for instance, be homaloidal or non-homaloidal . . . You cannot use the same coordinates for describing both kinds of space, any more than you can use rectangular coordinates on the surface of a sphere; that is, in fact, the geometrical interpretation of the difference. (Eddington, 1918, p. 23)

The question of "absoluteness," it would seem, is not for Eddington an ontological question of the sort that has historically been framed in the terms of the "absolute–relational" debate. It is a question of the empirical constraints on the kind of mathematical structure we may impose upon the world: the absolute properties of space-time are the recalcitrant properties that we confront when we wish to coordinatize it in a particular way. Einstein's remark that coordinates have no "immediate metrical significance," interpreted from Eddington's point of view, means that we can no longer extrapolate from a local coordinate system to the global structure of

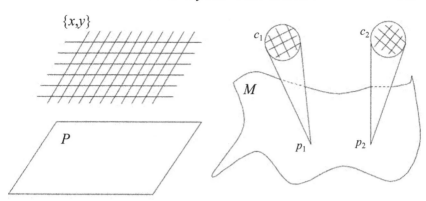

Figure 8. Coordinates in homaloidal and non-homaloidal spaces: the rectangular coordinate system $\{x, y\}$ can be "pasted" on to the plane P without distortion. But on the arbitrarily curved manifold M, no extended coordinate system can be smoothly applied; the spaces at distinct points p_1 and p_2 will have to be coordinatized in distinct local pieces or coordinate "patches" c_1 and c_2, which will in general be disoriented relative to one another.

space-time, or even to a finite region of space-time – not because space-time now has a different ontological status, but simply because it is curved. (See Figure 8.) So the physical interpretation of space-time geometry no longer corresponds to the idealized physical interpretation of coordinate systems, in which they were taken to be simply definable by the rigid displacement of rods and the uniform motion of clocks. Those physical processes are characteristic of flat spaces and space-times, but are ruled out by the presence of curvature. Even though coordinates are arbitrary, therefore, how and whether we may impose them depends on the physical characteristics of space-time, and hence is a matter of empirical fact.

The question of the arbitrariness of coordinates is closely related to the question of the arbitrariness of the gravitational field. Here again, however, Eddington had a compelling philosophical argument that there is a non-arbitrary feature of the world at stake. He certainly acknowledged the convenient practice of speaking of the metric tensor as the gravitational field, and the metric coefficients $g_{\mu\nu}$ as "gravitational potentials":

The double aspect of these coefficients, g_{11} &c., should be noted. (1) They express the metrical properties of the co-ordinates. This is the official standpoint of the principle of relativity, which scarcely recognizes the term "force." (2) They express the potentials of a field of force. This is the unofficial interpretation which we use when we want to translate our results in terms of more familiar conceptions. (Eddington, 1918, p. 23)

But he did offer what amounts to a philosophical defense of the first interpretation over the second. To understand this, it may be illuminating first to consider some critical remarks of Russell's on the matter. Comparing Eddington's view to Whitehead's defense of Euclidean geometry, Russell wrote: "Whereas Eddington seems to regard it as necessary to adopt Einstein's variable space, Whitehead regards it as necessary to reject it. For my part, I do not see why we should agree with either view: the matter seems to be one of convenience in the interpretation of formulae" (Russell, 1927, p. 78). As we have already seen, something like Russell's view was a central point in the logical positivist's interpretation of general relativity.

But Eddington in 1918 placed this issue in a rather different philosophical context. If we choose the geometrical over the force-field interpretation of gravity, it might seem as if we are hypothesizing about the nature of an unseen reality, in the face of facts that are compatible with both (or, in fact, any number of) hypotheses. But that would be a misunderstanding of the nature of spatio-temporal measurement, and its relation to the structure of space and time – as if the measurements were supposed to have any bearing at all on some metaphysical reality that is hidden behind them. Eddington's analysis bears quoting at some length:

The reader may not unnaturally suspect that there is an admixture of metaphysics in a theory which thus reduces the gravitational field to a modification of the metrical properties of space and time. This suspicion, however, is a complete misapprehension, due to the confusion of space, as we have defined it, with some transcendental and philosophical space. There is nothing metaphysical in the statement that under certain circumstances the measured circumference of a circle is less than π times the measured diameter; it is purely a matter for experiment. We have simply been studying the way in which physical measures of length and time fit together – just as Maxwell's equations describe how electrical and magnetic forces fit together. The trouble is that we have inherited a preconceived idea of the way in which measures, if "true," ought to fit. But the relativity standpoint is that we do not know, and do not care, whether the measures under discussion are "true" or not; and we certainly ought not to be accused of metaphysical speculation, since we confine ourselves to the geometry of measures which are strictly practical, if not strictly practicable. (Eddington, 1918, p. 29)

It would be easy to misread these remarks, as dissociating geometrical measurement from any concern with the real nature of space and time, that is, as a kind of positivistic dismissal of metaphysics. Their actual intent, however, is merely to question the existence of any distinction between "true" geometrical structure and the kind of structure that can conceivably be the object of our measurements. As long as we are not confused about such a distinction, the curvature of space-time is as much a matter of empirical measurement as ordinary spatial geometry ever was. More precisely, it is as measurable as the strength of the gravitational field ever was in the Newtonian setting, and measurable by the very same phenomena that we thought had been measuring the gravitational field – because, from the equivalence principle, we understand the relative accelerations of falling bodies as the relative divergence of local inertial frames. If there is another "true" geometrical structure underlying this one, in other words, the equivalence principle assures us that we can know nothing about it.

Eddington's discussion really addresses a more general problem concerning the relation between physics and ordinary experience, one closely analogous to the one faced by Helmholtz and his contemporaries concerning spatial geometry: namely, to justify construing some particular class of physical processes as standards for geometrical measurement. But the problem is transformed in the setting of a dynamical and variable space-time geometry. It is obvious enough that, especially in the case of general relativity, physics has separated the concept of space-time from our simple spatial "intuitions" about it. This has created a false impression, however, that in doing so physics has introduced strange hypotheses about the nature of space and time – and, therefore, that common sense has the right to ask whether space and time could really be as physics says they are. What this way of thinking overlooks is that relativity has not merely proposed to replace the intuitive notions of space and time with difficult theoretical notions. Rather, general relativity has subjected the intuitive notions to a kind of philosophical critique, and, in successive stages, brought them into closer harmony with physical knowledge of space-time measurement. The question that then emerges is, does common sense – even the common sense of the Newtonian physicist – have any clear conception of the structure of space-time independently of the measurements that physical laws make possible? In the nineteenth century, the question was whether we could form an empirical conception of spatial geometry other than the one exhibited by the dispositions of rigid bodies; after general relativity, the question was whether we could define a dynamical reference frame – and thereby identify the geodesics of space-time – other than by comparing the accelerations of falling bodies.

It is remarkable enough that, at a time when the dominant view of general relativity was Einstein's "Machian" view – and decades before any other view made its way into the mainstream of the philosophy of science – Weyl and Eddington brought to light both the mathematical structure of the theory, and its true philosophical relationship with its predecessors. Indeed, by doing the former as perspicuously as they did, they could hardly help doing the latter: it was obvious that Newtonian space-time, special relativity, and general relativity involved differing conceptions of the geometrical structure of the world, and of the physical processes that define it for us, rather than fundamentally differing views of the epistemology of geometry. Moreover, the work of Weyl and Eddington reveals the continuity between general relativity and its predecessors, not only from the metaphysical, but from a methodological point of view: as a theory of space-time structure as something that is empirically measurable. The logical positivists thought that the fundamental issue of space-time measurement was addressed by their discussion of point coincidences; further specification was left as a matter for convenient stipulations. But obviously the information that one might get from point coincidences hardly suffices to answer the kind of question that physics normally asks about the phenomena that general relativity is supposed to address: how are motions determined by physical fields? How is the motion of a planet determined by the mass of the Sun? How does the theoretical concept of curvature correspond to any measurable magnitude? So the importance of the "world-structure" view was not what it may have contributed to our understanding of the objective reality or "absoluteness" of space-time in the sense that is concerned in the absolute–relational debate. Its importance lay in recognizing questions about the structure of space-time as empirical questions. General relativity had not made the structure of space-time any less objective than it had ever been; rather, it revealed how the structure is contingent on empirical circumstances in ways that had never been imagined.

This discussion of Weyl and Eddington, and on their particular ways of representing and disseminating the content of general relativity, does not purport to be a thorough account of the acceptance of general relativity. Nor does it purport to establish their "world-structural" point of view as the correct or even as the dominant interpretation of general relativity. It is true that their view was maintained and advanced by notable figures such as Synge (1960), and that it later attained a degree of pre-eminence through classic geometrical accounts of relativity such as Misner *et al.* (1973) and Hawking and Ellis (1973). But that would surely not suffice to show that the "paradigm-shift" to general relativity was primarily driven by philosophical arguments like the ones I have presented. For my purposes, however, it is not really necessary to make such a case. In fact, the idea of a paradigm-shift to general relativity is in itself something of an exaggeration; the theoretical physics community as a whole seems to have maintained a certain reserve toward the theory for many decades, doubtless because of the paucity of very compelling empirical applications (see Eisenstadt, 1989). That reserve was much broken down by the extraordinary developments in empirical tests of general relativity beginning in the 1960s (see Will, 1993), but it would still be an exaggeration to say that Einstein's theory is the framework, in Kuhn's sense, that determines how physicists think about space, time, and gravitation. The expectation that the theory is only provisional, to be replaced eventually by a quantum theory of gravity, is too widespread for such a characterization to make much sense.

My aim, instead, here as in the earlier chapters, has been to make a philosophical point about the nature of conceptual transformation in the physics of space and time. A study like this one could hardly hope to answer the broader historical question, whether all or most of those who accepted general relativity were motivated by arguments like the ones I have presented. I have focused, therefore, on a more restricted set of philosophical questions: did Einstein have, after all, a legitimate philosophical argument for the theory? Was it a non-circular argument, that is, an argument that took the Newtonian theory as its starting point rather than (as Kuhn would suggest) assuming general relativity itself as its fundamental premise? Could it be recognized as such, not merely in hindsight, but by Einstein's own contemporaries? In other words, could such an argument *justify* a reconsideration of the fundamental structure of space-time, and exhibit a clear sense in which the reconsidered view really constitutes a deeper understanding of the phenomena that the old theory claimed to explain? I think that the foregoing account of Einstein's reasoning, and the discussion of Eddington and Weyl, suffice to answer these questions affirmatively.

4.7 THE PHILOSOPHICAL SIGNIFICANCE OF GENERAL RELATIVITY

In retrospect, the insistence on the relativity of motion, by Mach, Einstein, and their philosophical sympathizers, recalls the mechanical philosophers' insistence on reducing interaction to impact. On the one hand, in both cases a collection of philosophical ideas, both epistemological and metaphysical, provided a powerful motivation to free physics from certain traditional bonds, and to construct novel theories. On the other hand, the philosophical ideas themselves, in both cases, involved certain misconceptions. Under the guise of a sound methodological stricture against unintelligible "occult" causes, the mechanical philosophy proposed a purely hypothetical and, from the Newtonian point of view, more or less arbitrary restriction of all physical interaction to contact forces. Under the guise of a philosophical critique of an existing theory, on the grounds that it violates principles of empiricism and causal intelligibility, the relativist program really offered a metaphysical hypothesis: that the principle of inertia, as understood in Newtonian mechanics and special relativity, will prove to be reducible to some deeper kind of interaction that has not been hitherto understood. As in the case of the mechanists' demand for a mechanical model to explain the phenomena of universal gravitation, the mere possibility of imagining such a theory does not constitute a legitimate philosophical argument against the existing theory.[9] No more can the preference for the purportedly deeper theory claim some objective epistemological ground, since the theory is necessarily a speculative one, whereas the theory under attack – in this case, the principle of a privileged state of inertial motion – has empirically well-defined criteria of application. If anything, the problem was more acute for Einstein than for the mechanists since, Einstein's philosophical objections notwithstanding, he actually did succeed in defining a privileged state of motion through his application of the equivalence principle. For it was Einstein himself, after all, who identified the equivalence principle as the basis for a new definition of the space-time geodesic, and thereby an interpretation of the gravitational potential as an expression of the curvature of space-time. That this space-time theory emerged in spite of Einstein's philosophical objections, apparently, does not need to be explained by an appeal to some sort of mysterious physical intuition on Einstein's part. Rather, this circumstance exhibits Einstein's careful attention to the empirical meanings of the fundamental concepts that he employed – more precisely, to the empirical criteria for their application – and how little distracted his attention was from these problems, even by his most questionable philosophical ideas.

When we understand both the insufficiency of Einstein's preliminary philosophical motivations for the general theory of relativity, and the constructive significance of the equivalence principle, we can begin to understand how it is possible for Einstein to have created a theory that is so much at variance with his motives – how the theory turned out to have philosophical implications that disturbed him, and that were left for others to articulate clearly. The situation recalls a remark of Kant's, concerning synthetic and analytic definitions, and why the mathematical sciences have no use for analytic definitions:

> The general definition of similarity is of no concern whatever to the geometer. It is a fortunate thing for mathematics that, even though the geometer occasionally gets involved in the business of furnishing analytic definitions as a result of a false conception of his task, in fact nothing is actually inferred from such definitions, or, at any rate, the immediate inferences which he draws ultimately constitute the mathematical definition itself. Otherwise this science would be liable to exactly the same unfortunate discord as philosophy itself. (Kant, 1764, p. 277)

Einstein tried, on the basis of epistemological considerations borrowed from Mach, to articulate a general definition of motion as something purely relational, and of inertia as a kind of interaction with the contents of the universe at large. But the actual product of Einstein's reasoning – the general theory of relativity, the theory of curved space-time geometry – was not seriously affected by these considerations on the nature of motion, because the structure of the theory was determined in the end by a kind of synthetic principle, the theory of geodesic motion as the motion of freely falling bodies. This principle ensured the connection between the gravitational field and the curvature of space-time, and no amount of confusion about the theory's philosophical significance, or its relation to the ideas of Newton or Mach, could prevent the theory from developing as an empirical theory of physical geometry.

NOTES

1. See, for example, Stachel (2002a, 2002b), among other articles in Stachel (2002d).
2. For some further discussion of the development of the concept of inertial frame, see DiSalle (1990, 2002d). Regarding this development in relation to Mach, and Mach's critique of Newtonian mechanics, see DiSalle (2002c).
3. This is evidently an extremely brief and inadequate sketch of the geometry of Minkowski space-time. For a more detailed account see (for example) Geroch (1978), Taylor and Wheeler (1978).
4. In addition to Friedman (2002a), see Stachel (1989a) and Torretti (1983).
5. See Earman (1989, chapter 5) and Belot and Earman (2001).
6. For informative discussions of the historical development, see Stachel (1989b, 2002b) and Norton (1989a, b).
7. For a discussion of some of the serious interpretive issues that arise in the context of contemporary physics, particularly in connection with quantum theories of gravity, see Belot and Earman (2001).

8. In the literature of the philosophy of science, in fact, the serious re-assessment of Mach's view only began in the late 1960s, with works such as Stein (1967, 1977). See also DiSalle (2002b).
9. For further discussion of this issue, and the use of Machian ideas in twentieth-century and contemporary physics, see DiSalle (2002b). For some alternative views see, e.g., Barbour and Pfister (1995).

CHAPTER 5

Conclusion

5.1 SPACE AND TIME IN THE HISTORY OF PHYSICS

In the history of modern physics, space and time have after all played something like the role attributed to them by Kant. Not as forms of intuition: this was only incidentally the case, in a context where the geometry of space and the intuitive means of knowing about space seemed inseparable from one another. In that context, the processes of "representing to ourselves" in the productive imagination and of conceptualizing the relative situations of physical things appeared to be seamlessly connected. That is, the infinite Euclidean space in which physics treated the positions and motions of bodies was the most straightforward extension of the space in which we move, grasp our relation to our immediate surroundings, and situate our spatial point of view. But they have played the quasi-Kantian role of a framework that enables physics to constructively define its fundamental concepts of force and causality, by giving physics the means to construct such concepts as measurable theoretical quantities. The familiar and vague notion of force, through the work of Galileo, Huygens, Newton, and others, became a physical concept with a constructive spatio-temporal definition, one that did not really violate the common notion – even if it seemed to at first – but that rendered it a powerful tool of physical investigation, and thereby made the discovery of physical forces a clear and attainable goal. The connection with intuition was transformed by the recognition that intuition itself borrowed its self-evidence from elementary physical principles, principles so familiar as to be relied upon almost completely unconsciously. Then the true weak point of Kant's view was revealed: that the intuitive picture of space depended on principles that are physical and contingent, so that it could no longer be thought of as apodeictically certain or beyond revision; space could therefore no longer be a fixed framework for physics, but became something about which physics could eventually reveal surprising facts. Yet none of this fundamentally changed the role of

153

space and time as a framework for the construction of physical magnitudes with empirical measures, and for the posing of questions about force and causality as empirical questions. The framework remained indispensable to the scientific approach to nature in general, as Kant had aptly put it, "not as a pupil ready to accept whatever the teacher should recite, but as a judge compelling a witness to answer the questions that he sets" (Kant, 1787 [1956], p. Bxiii).

With the emergence of the notion of space-time, the connection with intuition may be said to have been dissolved altogether. The difference was not merely that physical principles could provide novel empirical facts about the nature of space, but that physics could cause us to reconsider the very principles by which we define spatial and temporal measurement. For the simple principle of rigid displacement that it had shared with spatial intuition, physics would substitute dynamical principles with no self-evident intuitive counterpart. Einstein's definition of simultaneity accorded well with the intuitive use of the concept, but to acknowledge it as the fundamental definition, rather than as just a practical substitute for absolute simultaneity, was to separate the objective features of space-time from everything that made spatial measurement seem intuitively evident. The laws of electrodynamics, essentially spatio-temporal in character, took precedence over the pre-theoretical notion of rigid spatial displacement, which consequently could reveal only a "complicated projection" of invariant geometrical relations on some arbitrary inertial frame. In retrospect, it emerged that it was only a simplistic assumption about simultaneity that made spatial relations appear so intuitively obvious in the first place. Yet even so the basic role of the spatio-temporal framework was not so radically transformed. Space-time, rather than space, was the framework within which physical magnitudes were to be constructed, and were understood as objectively meaningful to the extent that they corresponded to invariant features of the space-time structure.

Evidently this situation was altered by the emergence of general relativity; objective physical magnitudes evidently could not be defined as the invariants of a structure that would not, in general, have any large-scale symmetries. Yet it is important not to exaggerate the difference. The locally Minkowskian character of space-time implies that, at least at small scales and for short-range interactions, space-time remains a framework that allows for the construction of physical magnitudes and constrains the behavior of physical forces; the fact that gravitational effects may be set aside in such contexts means that this function of the space-time structure is, to a great degree, independent of the inhomogeneity of space-time at larger scales. That is only a straightforward application of the equivalence principle: it expresses in geometrical terms the fact that sufficiently small falling frames may be treated as inertial. It is a bizarre feature, perhaps, that a structure that plays this quasi-Kantian role, as the geometrical "form" within which the content of physical notions is defined, should also be a dynamical feature of the world, a field whose structure and evolution are determined by the contingent distribution of matter and energy. But it was, arguably, already bizarre in special relativity, that the geometry of space-time should turn out to depend upon a contingent fact about electromagnetic radiation; perhaps it was already bizarre about Newtonian mechanics, that acceleration of all things should emerge as the defining feature of force and therefore of the inertial structure of space-time. In fact what makes such developments seem bizarre is the older Kantian viewpoint, from which it seemed that such a form could be given to us prior to any physical principle, and from which the real interdependence of geometrical form and physical content was so difficult to see. Once that interdependence came to light, with the nineteenth-century recognition of the physical assumptions underlying geometrical intuition, a kind of dialectical engagement of physics and geometry began in earnest; its theoretical consequences have been incredibly far-reaching, and its end is nowhere in sight.

It will be clear now, I hope, from all that has been said, that the talk of dialectic is quite straightforward and unassuming, implying nothing particularly Hegelian – especially, nothing about the philosophical necessity or the historical inevitability of any of the developments I have discussed. On the contrary, what is dialectical about the history is also entirely contingent: for the revolutions we have considered, what has overthrown a given theory of space and time has been, not the "seeds of its own destruction" that it has carried internally, but the confrontation with unexpected contingent facts – facts which, on careful analysis, could be seen to undermine the concepts on which the theory had staked an entire spatio-temporal framework. The contradictions occur because it is the nature of such theories – as Newton often suggested, by word and example – to extrapolate far beyond the empirical evidence that originally motivated them, and so to *expose* themselves to contradictions that arise from unexpected empirical circumstances. Analyses like Einstein's, as we saw, of simultaneity and free-fall, might have been undertaken earlier. But at earlier stages in the development of physics, they could hardly have forced a re-evaluation of fundamental concepts; at the time when Einstein undertook those analyses, it was the current state of empirical knowledge that made them as consequential as they were. They took the form that they did, not because of the nature of theories, or of coordinative definitions, but because of the contingent nature of electromagnetism and gravity.

I have emphasized a dialectical element, then, chiefly because it seems to capture the peculiar combination of philosophical and empirical analysis by which novel space-time theories have emerged. Again, Kant and the logical positivists were right in a certain limited sense; the kind of a-priori principle that goes to constitute a spatio-temporal framework cannot be the same as an empirical principle, and cannot be justified by the usual sort of empirical or inductive argument – since empirical arguments, in the usual sense, must take such principles for granted. Indeed, this is part of their nature as principles for the interpretation of empirical facts. By their ways of understanding the interpretive character of the principles, however, both Kant and the positivists prevented themselves from seeing the role of contingent and empirical motivations. For Kant, the principles of Euclidean geometry and Newtonian physics were both necessary and sufficient for any understanding of physical phenomena as a genuine "world" of things in genuine physical interaction; there could be no contingent fact that could not be grasped within that framework. For the positivists, since interpretation required some arbitrary imposition of empirical content on a purely formal scheme, contingency could be appealed to only for the sake of pragmatic arguments about different possible conventions. One could say that in a different world, Newton's might have been a simpler and more useful space-time theory than Einstein's, but one could say little more than that. Along these lines, then, there was little hope of doing justice to the kind of argument that actually led Einstein to the theory, and enabled him to make a principled case for it to his contemporaries. The Newtonian principles could not be disproven by empirical or logical considerations alone, because they functioned as definitions of basic concepts rather than as empirical claims. Therefore they could confront the empirical facts, not inductively, but only dialectically.

Seeing the dialectical aspect of the arguments, moreover, illuminates the sense in which the history of space-time theory has been largely a progressive development. Kuhn and the positivists were undoubtedly right to deny that the changes have been cumulative: given the radical conceptual transformations they have required, the successive theories can hardly be called mere additions to existing knowledge. But for the positivists this meant that the new theories met epistemological standards that the older ones had failed, while for Kuhn it meant that the theories were altogether incommensurable. Neither view puts the historical development in its proper philosophical light. The kind of progress that each theory has offered, we have seen, is an enlargement of perspective: a philosophical analysis of existing conceptions and their empirical foundations, and a more comprehensive viewpoint from which those conceptions are revealed to be local and incomplete. Special relativity "contains" Newtonian space-time, for example, not merely by making the same predictions in a limiting case, but by revealing that what passes for absolute simultaneity is in fact a narrow and relative conception, mistakenly extended from a particular frame of reference to every inertial frame. Similarly, the argument for general relativity reveals the conception of inertial frame to be a purely local one, mistakenly extended to the global structure of space-time. As a result of such conceptual changes, physicists might well consider themselves to be living in a different "world." But the path to that world begins with a conceptual analysis of geometrical concepts in the familiar world, and ends in a perspective from which the familiar concepts have a natural place within a more comprehensive framework, in which facts that had seemed contradictory now form a coherent whole. The result of such a transformation is, simply, a deeper understanding of the nature of space and time.

This account makes it possible, finally, to rehabilitate a central idea of the logical positivists: that there were philosophical arguments for special and general relativity, and that creating the theories was as much a work of philosophical analysis as of scientific discovery. In the form that the logical positivists gave it, that idea was discredited, because it seemed to rest on extremely simplistic philosophical notions regarding metaphysics, meaning, and the relation between theory and observation. But now we can see that the philosophical analysis of space and time has been, at least in the cases that matter, something more subtle than the mere application of epistemological strictures or slogans. Despite the delusions of philosophers and scientists of having purely epistemological or metaphysical insights into the nature of space, time, and motion, philosophy is not an independent source of knowledge of space-time; our ability to conceive of or to reason about space has always depended on principles borrowed, explicitly or implicitly, from physics. But this is not to say that physics simply provides answers to philosophical questions from its own resources, or that philosophy has to content itself with accepting them. Rather, it says that, at certain critical points in its history, the fundamental problems of physics have to do with the ways in which fundamental concepts are defined. In those circumstances, the pursuit of physics in accord with those concepts evidently has not resolved the underlying problems. These are the times at which philosophical analysis has become an unavoidable task for physics itself.

5.2 ON PHYSICAL THEORY AND INTERPRETATION

This book suggests an alternative, not only to the logical positivists' account of the history of physics, but also to their view of the nature of theories. It suggests that the picture of scientific theories as uninterpreted formal systems, linked to experience by arbitrary stipulations, involves some deep misunderstandings about the nature of interpretive principles – and, there-fore, of the nature and evolution of scientific theories. It is natural that subsequent philosophers should have sought some other way of under-standing the connections between physical theories and the phenomena that they are supposed to explain. But the post-positivist tendency in the philosophy of science has been not to seek a better account of the interpre-tation of scientific theories, but to set aside the problem of interpretation altogether. On the "semantic view of theories," a physical theory is not considered "syntactically" as an axiomatic system combined with a set of interpretive rules (such as the "coordinative definitions" emphasized by the logical positivists). Rather, it is considered in the now-familiar model-theoretic terms, that is, as a structure with a set of models. The genuine differences between the "semantic view" and the positivists' view have to do with the differences between semantic and syntactic conceptions of structure in general – not with any philosophical difference concerning the way in which a structure can be applied to experience. That a given structure has "the world" as one of its models is typically represented as an "empirical hypothesis." The meaning of this claim is not an object of any serious philosophical scrutiny; it is simply taken for granted that the structure has a natural or "intended" interpretation. But it is no less true on the semantic view than on the syntactic view that, say, Euclidean geometry might just as well be the structure of a universe of "tables, chairs, and beer mugs"; hence either the claim that "the world" has that structure is an entirely trivial one, or some serious examination of the meaning of that claim is urgently required (see Demopoulos, 2003). Indeed, it is not merely the issues surrounding conventionalism, but any number of other issues that are still taken seriously in the philosophy of science – issues of real-ism and antirealism, objectivism and relativism, and the general question of the rationality of science itself – that require some discussion of how scientific theories, more than other symbolic interpretive frameworks or "belief-systems" or "ways of knowing," can manage to make meaningful claims about the nature of things.

It was not without reason, then, that philosophers such as Kant and the logical positivists hoped that, given a persuasive account of how science generates meaningful statements, issues regarding ontology and rationality could fade into insignificance. Convinced that science was not inherently more rational than other intellectual pursuits, especially metaphysics, and was no better able to discern the real ontology (the nature of "things in themselves") underlying the phenomenal world, they credited mathematical physics with a just sense of these limitations, and an implicit grasp of the need to impose some structure on the phenomena – an understanding that the phenomena only constitute a "world" to the extent that we can frame them in some systematic interconnection. It was implicitly understood that science could not address previously defined metaphysical questions, or any purely philosophical question about "what there is," because general metaphysics had never posed such questions in any answerable form. For that reason metaphysics was doomed to endless controversy between answers that could claim no more than a subjective plausibility. Physics, meanwhile had imposed a conception on the phenomenal world in virtue of which "what there is" could become an empirical question. For Kant, what exists is what can be situated in the framework of Euclidean space and Newtonian time, and can be seen to stand in causal interrelationships according to the causal principles defined by Newtonian physics. Traditional metaphysics might dispute the right of physics to restrict the question in this way, but it had no convincing alternative way of specifying the question – at least, none that did not implicitly borrow some of its essential content from assumptions about space and time. This is why the status of space and time could not be, for Kant, the sort of ontological issue it was for the Newtonians and Leibnizians, or became in the later twentieth century. It was an issue concerning physics' need for a framework within which concepts of substance, force, and causality could be physically meaningful, and play essential roles in a true metaphysics of nature. It was an issue for transcendental analysis, not for the endless debate between rival metaphysical hypotheses.

For the logical positivists, the framework that Kant had thought both sufficient and necessary was revealed to be neither: newer physical theories could comprehend phenomena for which the Newtonian framework was inadequate, and the multiplicity of possible such theories meant that there could be no question of necessity. On the contrary, the assignment of an interpretive framework to the phenomena involved a degree of arbitrariness that Kant, for whom the intuitive interpretation of geometry was unique and beyond doubt, could not have imagined. Hence the resort to conventionalism, and the conviction that the adoption of any particular interpretive framework must be a matter for pragmatic negotiation rather than theoretical reasoning. Yet their approach to metaphysics and ontology was, in spirit, the same. Frameworks might be arbitrary, but, at least, within their confines, questions about the existence, the nature, and the interconnections of physical things could be so posed that empirical evidence could, in principle, answer them. Outside of such a framework, however, such questions could have no real meaning at all, and indeed were properly understood as pseudo-questions. Unless it was restricted to the internal ontology fixed by some interpretive framework, metaphysics could only mean a kind of hopeless effort: the effort to answer questions like "what there is," in a context removed from all possible logical or empirical means of answering it. For the positivists as for Kant, in short, true science was distinguished not by rationality, or by rational insight into what lies behind the phenomena, but by knowing what it is talking about – by knowing what questions it can meaningfully ask, and knowing how to judge whether it has found an answer. Kant and the positivists thus extended a thought that was always part of classical empiricism, for example in the thought of Berkeley and Hume: that traditional metaphysical controversy takes the question "what is real and what is illusion?," which makes sense in a certain empirical context, and tries to ask it in a setting in which it makes no sense at all, as a question about the empirical world as a whole – "does the world as it appears resemble the world as it really is?" So traditional metaphysics failed to see that the empirical world itself is the only framework within which such questions can be meaningfully posed. But Kant and the positivists saw that the formal principles of science, and especially the principles of space and time, play a more essential role in defining that framework than Berkeley or Hume had ever imagined.

   In fact it is not too much to say that the division between structure and interpretation is, in itself, the greatest obstacle to a clear understanding of the way in which physical theories confront the world of experience. When we ask how the principles of a theory are to be interpreted, or how the structure associated with a theory is to be interpreted, we have already lost sight of the genuine content of those principles. For the principles are not, after all, purely formal principles in need of interpretation; rather, they are themselves principles of interpretation. Newton's laws, for example, do not constitute an empty formal structure; rather they constitute an interpretation of the phenomena of motion – more precisely, they constitute a program to interpret all accelerations as revealing the interplay of forces. This is not changed by the fact that we can present the laws in a seemingly abstract way, without considering any particular physical situations or genuine empirical cases. Such a presentation may be "sterile," as Newton says of the proofs in Book I of the *Principia* (1726 [1999], p. 793), but it does not yield an uninterpreted calculus. On the contrary, it is merely the demonstration of the laws' interpretive scope and power – a demonstration of everything that may follow from interpreting accelerations in this way. What follows, above all, is that every acceleration we observe may, at least in principle, reveal something about the nature, the origins, and especially the magnitudes of the physical forces at work.

Einstein's definition of simultaneity, similarly, is not a coordinative definition for some mathematical object; it is, rather, an interpretation of simultaneity, an attempt to articulate a conception of simultaneity that reveals its empirical meaning. That empirical meaning, moreover, is not merely its translation in an "observation language" or its operational definition, but its interconnections with the empirical principles that we rely upon in theoretical physics. Einstein's use of the equivalence principle is not merely a coordinative definition for the geodesics of an arbitrary Riemannian manifold; it is an interpretation of the geodesics of space-time. The question that it begins with is not, what physical significance should we attach to this mathematical object, the geodesic? The question is, rather, how can we distinguish any given motion as a geodesic of space-time? Or, how can we complete the Newtonian project of decomposing an accelerated motion into its inertial and gravitational parts? This interpretive aspect of the laws of physics is the source of their a-priori and seemingly unrevisable character; their actual revisability reflects what a stringent requirement it is upon such a theory, that it be capable of bringing the relevant phenomena within its interpretive grasp. Kant had understood this latter point, but the prospect of revision was one that he did not take seriously – precisely because he understood that Newton's was the only set of principles that had ever provided an interpretation of motions in this stringent sense of the term, and he was unable to conceive of phenomena that those principles could not eventually grasp.

All of this suggests that there is a great deal of truth in the remark that modern physics, under the influence of Newton, has had to create "its own theory of measurement" (Smith, 2003a). But for the physics of space and time, perhaps this point should be stated even more strongly: in a certain sense, space-time physics *is* its theory of measurement; it is a program to interpret certain characteristic phenomena as measurements of fundamental dynamical quantities, and then, to interpret mathematical relations among the quantities as expressing physical relations among the phenomena. I don't believe that this last point is affected by the possibility that, in our own time, research into quantum gravity is likely to yield a replacement for general relativity – not only that, but a theory in which space-time theory as Newton and Einstein understood it will no longer be fundamental, and some other kind of structure will play the fundamental role in that theory that space-time has played up to now. If philosophers and physicists are to make philosophical sense of such a structure, surely they will require a clear understanding – clearer, at any rate, than twentieth-century philosophy of science was able to achieve – of what the role of space-time structure really was, and how it functioned as a framework for other physical objects, interactions, and processes. I hope that this book has been a step toward that understanding.

References

Barbour, J. and Pfister, H. (eds) (1995). *Mach's Principle: From Newton's Bucket to Quantum Gravity.* Einstein Studies, vol. 6. Boston: Birkhäuser.

Belot, G. and Earman, J. (2001). Pre-Socratic quantum gravity. In *Physics Meets Philosophy at the Planck Scale,* eds C. Callander and N. Huggett. Cambridge: Cambridge University Press, pp. 213–55.

Ben-Menachem, Y. (2001). Convention: Poincaré and some of his critics. *British Journal for the Philosophy of Science,* **52**, 471–513.

Bishop, R. and Goldberg, S. (1980). *Tensor Analysis on Manifolds.* New York: Dover Publications.

Bolzano, B. (1817). Rein analytische Beweis des Lehrsatz. In *B. Bolzano, Early Mathematical Works, 1781–1848,* ed. L. Novy. Prague: Institute of Slovak and General History, 1981.

Carnap, R. (1956).Empiricism, semantics, and ontology. In *Meaning and Necessity.* Chicago: University of Chicago Press, Supplement A, pp. 205–21.

(1995). *An Introduction to the Philosophy of Science.* New York: Dover Publications (reprint).

Carrier, M. (1994). Geometric facts and geometric theory: Helmholtz and 20th-century philosophy of physical geometry. In *Universalgenie Helmholtz: Rückblick nach 100 Jahren,* ed. L. Kruger. Berlin: Akademie-Verlag.

Coffa, J. A. (1983). From geometry to tolerance: sources of conventionalism in the 19th century. In *From Quarks to Quasars,* ed. R. G. Colodny. *Pittsburgh Studies in the Philosophy of Science,* vol. X. Pittsburgh: University of Pittsburgh Press.

(1991). *The Semantic Tradition from Kant to Carnap.* Cambridge: Cambridge University Press.

Demopoulos, W. (2000). On the origin and status of our conception of number. *Notre Dame Journal of Formal Logic,* **41**, 210–26.

(2003). On the rational reconstruction of our theoretical knowledge. *British Journal for the Philosophy of Science,* **54**, 371–403.

Descartes, R. (1983). *The Principles of Philosophy,* transl. V. R. Miller and R. P. Miller. Dordrecht: Reidel.

Dingler, H. (1934). H. Helmholtz und die Grundlagen der Geometrie. *Zeitschrift für Physik,* **90**, 348–54.

DiSalle, R. (1990). The "essential properties" of matter, space, and time. In *Philosophical Perspectives on Newtonian Science*, eds P. Bricker and R. I. G. Hughes. Cambridge, MA: MIT Press.

(1991). Conventionalism and the origins of the inertial frame concept. In *PSA 1990: Proceedings of the 1990 Biennial Meeting of the Philosophy of Science Association*. East Lansing: The Philosophy of Science Association.

(2002a). Newton's philosophical analysis of space and time. In *The Cambridge Companion to Newton*, eds I. B. Cohen and G. E. Smith. Cambridge: Cambridge University Press.

(2002b). Conventionalism and modern physics: a re-assessment. *Noûs*, **36**, 169–200.

(2002c). Reconsidering Ernst Mach on space, time, and motion. In *Reading Natural Philosophy: Essays in the History and Philosophy of Science and Mathematics to Honor Howard Stein on his 70th Birthday*, ed. D. Malament. Chicago: Open Court Press.

(2002d). Space and time: inertial frames. In *The Stanford Encyclopedia of Philosophy*, <http://plato.stanford.edu/archives/win2003/entries/spacetime-iframes/>.

(2006). Kant, Helmholtz, and the meaning of empiricism. In *Kant's Legacy*, eds M. Friedman and A. Nordmann. Cambridge, MA: MIT Press.

Earman, J. (1989). *World Enough and Spacetime: Absolute and Relational Theories of Motion*. Cambridge, MA: MIT Press.

Eddington, A. S. (1918). *Report on the Relativity Theory of Gravitation*. London: Fleetwood Press.

(1920). *Space, Time, and Gravitation. An Outline of General Relativity Theory*. Cambridge: Cambridge University Press.

(1923). *The Mathematical Theory of Relativity*. Cambridge: Cambridge University Press.

Ehlers, J. (1973a). The nature and structure of space-time. In *The Physicist's Conception of Nature*, ed. J. Mehra. Dordrecht: Reidel, pp. 71–95.

(1973b). A survey of general relativity theory. In *Relativity, Astrophysics, and Cosmology*, ed. W. Israel. Dordrecht: Reidel.

Einstein, A. (1905). Zur elektrodynamik bewegter Körper. *Annalen der Physik*, **17**, 891–921.

(1911). On the influence of gravitation on the propagation of light. In *The Principle of Relativity*, eds A. Einstein, H. A. Lorentz, H. Minkowski and H. Weyl, transl. W. Perrett and G. B. Jeffery. New York: Dover Books, 1952, pp. 97–108.

(1916). *Die Grundlage der allegemeinen Relativitätstheorie*. Leipzig: Johann Ambrosius Barth. (Reprint from *Annalen der Physik*, (4) **49**, 769–822.)

(1917). *Über die spezielle und die allgemeine Relativitätstheorie (Gemeinverständlich)*, 2nd edn. Braunschweig: Vieweg und Sohn.

(1919). Was ist Relativitäts-Theorie? In *The Collected Papers of Albert Einstein*, vol. 7, eds M. Jansen, R. Shulmann, J. Illy, C. Lehner and D. Buchwald. Princeton, NJ: Princeton University Press, pp. 206–11.

(1920). Grundgedanken und Methoden der Relativitätstheorie in ihrer Entwick-elung dargestellt. In *The Collected Papers of Albert Einstein*, vol. 7, eds M. Jansen, R. Shulmann, J. Illy, C. Lehner and D. Buchwald. Princeton, NJ: Princeton University Press, pp. 212–49.

(1922). *The Meaning of Relativity*. Princeton, NJ: Princeton University Press.

(1949). Autobiographical notes. In *Albert Einstein, Philosopher-Scientist*, ed. P. A. Schilpp. Chicago: Open Court, pp. 2–94.

Eisenstadt, J. (1989). The low-water mark of general relativity, 1925–1950. In *Einstein and the History of General Relativity*, Einstein Studies, vol. 1, eds D. Howard and J. Stachel. Boston: Birkhäuser, pp. 277–92.

Euler, L. (1748). Réflexions sur l'espace et le temps. *Histoire de l'Academie Royale des sciences et belles lettres*, **4**, 324–33.

(1765). *Theoria motus corporum solidorum*. Rostock and Greifswald, 1765.

Flores, F. (1999). Einstein's theory of theories and types of theoretical explanation. *International Studies in the Philosophy of Science*, **13**, 123–34.

Fock, V. (1959). *The Theory of Space, Time, and Gravitation*, transl. N. Kemmer. London: Pergamon Press.

Friedman, M. (1983). *Foundations of Space-time Theories*. Princeton, NJ: Princeton University Press.

(1990). Kant and Newton: why gravity is essential to matter. In *Philosophical Perspectives on Newtonian Science*, eds P. Bricker and R. I. G. Hughes. Cambridge, MA: MIT Press.

(1992). *Kant and the Exact Sciences*. Cambridge, MA: Harvard University Press.

(1999a). Geometry, convention, and the relativized a priori: Reichenbach, Schlick, and Carnap. In *Reconsidering Logical Positivism*. Cambridge: Cambridge University Press, pp. 59–70.

(1999b). Poincaré's conventionalism and the logical positivists. In *Reconsidering Logical Positivism*. Cambridge: Cambridge University Press, pp. 71–86.

(1999c). Geometry, construction, and intuition in Kant and his successors. In *Between Logic and Intuition: Essays in Honor of Charles Parsons*, eds G. Scher and R. Tieszen. Cambridge: Cambridge University Press.

(2002a). Geometry as a branch of physics: background and context for Einstein's "Geometry and Experience". In *Reading Natural Philosophy: Essays in the History and Philosophy of Science and Mathematics to Honor Howard Stein on his 70th Birthday*, ed. D. Malament. Chicago: Open Court Press.

(2002b). *The Dynamics of Reason: the 1999 Kant Lectures at Stanford University*. Chicago: University of Chicago Press.

Galileo (1632 [1996]). *Dialogo Sopra I Due Massimi Sistemi del Mondo – Ptolemaico e Copernicano*. Florence, 1632. Reprint, Milan: Oscar Mondadori.

Geroch, R. (1978). *General Relativity from A to B*. Chicago: University of Chicago Press.

Hall, A. R. and Hall, M. B. (eds) (1962). *Unpublished Scientific Papers of Isaac Newton*. Cambridge: Cambridge University Press.

Hawking, S. and Ellis, G. F. R. (1973). *The Large-Scale Structure of Space-Time*. Cambridge: Cambridge University Press.

Helmholtz, H. (1868). Über die Thatsachen, die der Geometrie zum Grunde liegen. *Nachrichten von der königlichen Gesellschaft der Wissenschaften zu Göttingen*, **9**, 193–221. Reprinted in *Wissenschaftliche Abhandlungen*, **2**, 618–39.

    (1870). Ueber den Ursprung und die Bedeutung der geometrischen Axiome. In *Vorträge und Reden*, 2 vols. Braunschweig: Vieweg und Sohn, pp. 1–31.

    (1878). Die Thatsachen in der Wahrnehmung. In *Vorträge und Reden*, 2 vols. Braunschweig: Vieweg und Sohn, pp. 215–47.

    (1887). Zählen und Messen, erkenntnisstheoretische betrachtet. *Wissenschaftliche Abhandlungen*, vol. 3. Leipzig: J. A. Barth, pp. 356–91.

    (1921). *Schriften zur Erkenntnistheorie*, eds P. Hertz and M. Schlick. Berlin: Springer-Verlag.

Hughes, R. I. G. (1987). *The Structure and Interpretation of Quantum Mechanics*. Cambridge: Cambridge University Press.

Kant, I. (1764 [1911]). Untersuchung ueber die Deutlichkeit der Grundsaetze der naturlichen Theologie und der Moral (the "Prize Essay"). In *Gesammelte Schriften*. Akademie Ausgabe, Berlin: Georg Reimer, vol. 2, pp. 273–301.

    (1768 [1911]). Von dem ersten Grunde des Unterschiedes der Gegenden im Raume. In *Gesammelte Schriften*. Akademie Ausgabe, Berlin: Georg Reimer, vol. 2, pp. 375–83.

    (1770). De mundi sensibilis atque intelligibilis forma et principiis. In *Gesammelte Schriften*. Akademie Ausgabe, Berlin: Georg Reimer, vol. 2, pp. 385–419.

    (1783). Prolegomena zu einer jeden künftigen Metaphysik die als Wissenschaft wird auftreten können. In *Gesammelte Schriften*. Akademie Ausgabe, Berlin: Georg Reimer, vol. 4.

    (1786 [1911]). *Metaphysische Anfangsgründe der Naturwissenschaft*. In *Gesammelte Schriften*. Akademie Ausgabe, Berlin: Georg Reiner, vol. 4, pp. 465–565.

    (1787 [1956]). *Kritik der reinen Vernunft*. Reprint, Berlin: Felix Meiner Verlag.

Klein, F. (1872). *Vergleichende Betrachtungen über neuere geometrische Forschungen*. Erlangen: A. Duchert.

Kretschmann, E. (1917). Ueber die physikalischen Sinn der Relativitätspostulaten. *Annalen der Physik*, (4) **53**, 575-614.

Kuhn, T. (1970a). *The Structure of Scientific Revolutions*, 2nd edn. Chicago: University of Chicago Press.

    (1970b). Logic of discovery or psychology of research? In *Criticism and the Growth of Knowledge*, eds I. Lakatos and A. Musgrave. Cambridge: Cambridge University Press.

    (1977). A function for thought-experiments. In *The Essential Tension*. Chicago: University of Chicago Press.

Lange, L. (1885). Ueber das Beharrungsgesetz. *Berichte der Königlichen Sachsischen Gesellschaft der Wissenschaften zu Leipzig, Mathematisch-physische Classe*, **37**, 333–51.

Leibniz, G. W. (1694). Letter to C. Huygens. In *Die mathematische Schriften von Gottfried Wilhelm Leibniz*. Berlin, 1849–55. Reprint, Hildeshein: Georg Olms, vol. II, pp. 179–85.

(1695 [1960]). Systeme nouveau de la nature et de la communication des substances, aussi bien que l'union qu'il y a entre l'ame le corps. In *Die philosophischen Schriften von Gottfried Wilhelm Leibnitz*. Berlin, 1875–90. Reprint, Hildeshein: Georg Olms, vol. IV, pp. 477–87.

(1699). Letter to B. de Volder. In *Die philosophischen Schriften von Gottfried Wilhelm Leibnitz*. Berlin, 1875-90. Reprint, Hildeshein: Georg Olms, pp. 168–70.

(1716). Correspondence with S. Clarke. In *Die philosophischen Schriften von Gottfried Wilhelm Leibnitz*. Berlin, 1875-90. Reprint, Hildeshein: Georg Olms, vol. VII, pp. 345–440.

Lorentz, H. A. (1895). Michelson's interference experiment. In *The Principle of Relativity*, eds A. Einstein, H. A. Lorentz, H. Minkowski and H. Weyl, transl. W. Perrett and G. B. Jeffery. New York: Dover Books, pp. 3–7.

(1904). Electromagnetic phenomena in a system moving with any velocity less than that of light. In *The Principle of Relativity*, eds A. Einstein, H. A. Lorentz, H. Minkowski and H. Weyl, transl. W. Perrett and G. B. Jeffery. New York: Dover Books, pp. 11–34.

Mach, E. (1883). *Die Mechanik in ihrer Entwickelung, historisch-kritisch dargestellt.* Leipzig: Brockhaus.

(1889). *Die Mechanik in ihrer Entwickelung, historisch-kritisch dargestellt*, 2nd edn. Leipzig: Brockhaus.

Magnani, L. (2002). *Philosophy and Geometry: Theoretical and Historical Issues.* Western Ontario Series in Philosophy of Science, vol. 66. Dordrecht: Kluwer.

Malament, D. (1986). Newtonian gravity, limits, and the geometry of space. In *From Quarks to Quasars: Philosophical Problems of Modern Physics*, ed. R. Colodny. Pittsburgh: Pittsburgh University Press.

Maxwell, J. (1877). *Matter and Motion*. New York: Dover Publications (reprint 1952).

Mill, J. S. (1843). *A System of Logic.* London: Parker and Son.

Minkowski, H. (1908). Die Grundgleichungen für die elektromagnetischen Vorgänge in bewegten Körper. *Nachrichten der königlichen Gesellschaft der Wissenschaften zu Göttingen, mathematisch-physische Klasse*, pp. 53–111.

(1909). Raum und Zeit. *Physikalische Zeitschrift*, **10**, 104–11.

Misner, C., Thorne, K. and Wheeler, J. A. (1973). *Gravitation*. New York: W. H. Freeman.

Nagel, E. (1939). The formation of modern conceptions of formal logic in the development of geometry. *Osiris*, **7**, 142–224.

Neumann, C. (1870). *Ueber die Principien der Galilei-Newton'schen Theorie.* Leipzig: B. G. Teubner.

Newcombe, S. (1910). Light. In *Encyclopaedia Britannica*, 11th edn, vol. **16**, sect. III, pp. 623–6.

Newton, I. (1704 [1952]). *Opticks.* London. Reprint, New York: Dover Publications.

168   *References*

(1726 [1999]). *The Principia: Mathematical Principles of Natural Philosophy*, transl. I. B. Cohen and A. Whitman. Berkeley and Los Angeles: University of California Press.

(1729 [1962]). *The System of the World.* In *Sir Isaac Newton's Mathematical Principles of Natural Philosophy and his System of the World,* ed. F. Cajori, transl. A. Motte, 2 vols. Berkeley: University of California Press.

Norton, J. (1989a). What was Einstein's principle of equivalence? In *Einstein and the History of General Relativity*. Einstein Studies, vol. 1, eds D. Howard and J. Stachel. Boston: Birkhäuser, pp. 5–47.

(1989b). How Einstein found his field equations. In *Einstein and the History of General Relativity*. Einstein Studies, vol. 1, eds D. Howard and J. Stachel. Boston: Birkhäuser, pp. 101–59.

Poincaré, H. (1899a). Des fondements de la géométrie; a propos d'un livre de M. Russell. *Revue de Metaphysique et de Morale*, **VII**, 251–79.

(1899b). Des fondements de la géométrie; réponse à M. Russell. *Revue de Metaphysique et de Morale*, **VIII**, 73–86.

(1902). *La Science et L'Hypothèse.* Paris: Flammarion.

(1905). Sur la dynamique de l'électron. *Comptes rendues de l'Académie des Sciences*, **140**, 1504–8.

(1913). *Dernières Pensées.* Paris: Flammarion.

Quine, W. V. O. (1953). Two dogmas of empiricism. In *From a Logical Point of View*. New York: Harper, pp. 20–46.

Reichenbach, H. (1924). Die Bewegungslehre bei Newton, Leibniz, und Huyghens. *Kantstudien*, **29**, 239–45.

(1949). The philosophical significance of relativity. In *Albert Einstein, Philosopher-Scientist*, ed. P. A. Schilpp. Chicago: Open Court, pp. 289–311.

(1957). *The Philosophy of Space and Time*, transl. M. Reichenbach. New York: Dover Publications. (Originally published as *Philosophie der Raum-Zeit-Lehre*, Berlin, 1927.)

Riemann, B. (1867). Ueber die Hypothesen, die der Geometrie zu Grunde liegen. In *The Collected Works of Bernhard Riemann*, ed. H. Weber. Leipzig: B. G. Teubner, 1902, pp. 272–87. Reprint, New York: Dover Publications, 1956.

Russell, B. (1897). *An Essay on the Foundations of Geometry*. Cambridge: Cambridge University Press.

(1899). Sur les axiomes de la géométrie. *Revue de Metaphysique et de Morale*, **VII**, 684–707.

(1927). *The Analysis of Matter*. Cambridge: Cambridge University Press.

Schlick, M. (1917). *Raum und Zeit in der gegenwärtigen Physik. Zur Einführung in das Verständnis der Relativitäts- und Gravitationstheorie*. Berlin.

Sklar, L. (1977). *Space, Time and Spacetime*. Berkeley, CA: University of California Press.

Smith, G. E. (2003a). How Newton's *Principia* changed physics. Unpublished manuscript.

(2003b). *Newton's Principia.* Unpublished lecture notes, Tufts University.

Spivak, M. (1967). *A Comprehensive Introduction to Differential Geometry*. Berkeley, CA: Publish or Perish Press.

Stachel, J. (1989a). The rigidly rotating disk as the "missing link" in the history of general relativity. In *Einstein and the History of General Relativity*. Einstein Studies, vol. 1, eds D. Howard and J. Stachel. Boston: Birkhäuser, pp. 48–62.

    (1989b). Einstein's search for general covariance. In *Einstein and the History of General Relativity*. Einstein Studies, vol. 1, eds D. Howard and J. Stachel. Boston: Birkhäuser, pp. 63–100.

    (2002a). "What song the sirens sang": How did Einstein discover special relativity? In *Einstein from B to Z*. Einstein Studies, vol. 9. Boston: Birkhäuser, pp. 157–70.

    (2002b). The genesis of general relativity. In *Einstein from B to Z*. Einstein Studies, vol. 9. Boston: Birkhäuser, pp. 233-4.

    (2002c). Einstein and Newton. In *Einstein from B to Z*. Einstein Studies, vol. 9. Boston: Birkhäuser, pp. 447-52.

    (2002d). *Einstein from B to Z*. Einstein Studies, vol. 9. Boston: Birkhäuser.

Stein, H. (1967). Newtonian space-time. *Texas Quarterly*, **10**, 174–200.

    (1977). Some philosophical prehistory of general relativity. In *Foundations of Space-Time Theories*, Minnesota Studies in Philosophy of Science, vol. 8, eds J. Earman, C. Glymour and J. Stachel. Minneapolis: University of Minnesota Press, pp. 3–49.

    (2002). Newton's metaphysics. In *The Cambridge Companion to Newton*, eds I. B. Cohen and G. E. Smith. Cambridge: Cambridge University Press.

Synge, J. L. (1960). *Relativity: The General Theory*. Amsterdam: North-Holland.

Taylor, E. and Wheeler, J. A. (1978). *Spacetime Physics*. New York: Wiley.

Thomson, J. (1884). On the law of inertia; the principle of chronometry; and the principle of absolute clinural rest, and of absolute rotation. *Proceedings of the Royal Society of Edinburgh*, **12**, 568–78.

Torretti, R. (1977). *Philosophy of Geometry from Riemann to Poincaré*. Dordrecht: Riedel.

    (1983). *Relativity and Geometry*. Oxford: Pergamon Press.

    (1989). *Creative Understanding*. Chicago: University of Chicago Press.

Trautman, A. (1965). Foundations and current problems of general relativity. In *Lectures on General Relativity*. Brandeis 1964 Summer Institute on Theoretical Physics, vol. 1, eds A. Trautman, F. A. E. Pirani and H. Bondi. Englewood Cliffs, NJ: Prentice-Hall.

    (1966). The general theory of relativity. *Soviet Physics Uspekhi*, **89**, 319–39.

Truesdell, C. (1967). Reactions of late Baroque mechanics to success, conjecture, error, and failure in Newton's *Principia*. *Texas Quarterly*, **10**, 238-58.

Van Fraassen, B. (1989). *Laws and Symmetries*. Oxford: Oxford University Press.

Weyl, H. (1918). *Raum-Zeit-Materie. Vorlesung über allgemeine Relativitätstheorie*. Berlin: Springer-Verlag.

(1927). *Philosophie der Mathematik und der Naturwissenschaften.* In Oldenburg's *Handbuch der Philosophie.* Munich and Berlin: Verlag R. Oldenburg.

Will, C. (1993). *Theory and Experiment in Gravitational Physics,* revised edn. Cambridge: Cambridge University Press.

Wilson, C. (2002). Newton and celestial mechanics. In *The Cambridge Companion to Newton,* eds I. B. Cohen and G. E. Smith. Cambridge: Cambridge University Press.

Index

absolute space   Section 2.5 *passim*, Section 3.4
    passim
    inertial frames as replacement for   28–30
    Leibniz's critique of   6, 14, 26–7, 52–3
    Mach's critique of   14, 34–5, 52, 53, 133–4
    Newton's definition of   17–18, 19–20,
        Figure 3
    Newton's arguments for   16–17, 37–8,
        48–50
    relationalist arguments against   13–14, 52–3
absolute time
    Einstein's critique of   Section 4.2 *passim*,
        9, 102–3, 109–11
    Leibniz's critique of   20, 22, 38, 52–3
    Mach's critique of   22–4, 52
    Newton's definition of   17–18, 19–20, 49,
        Figure 1
    Newton's arguments for   Section 2.4 *passim*,
        16–17, 20–21, 38, 49
    role in Newtonian mechanics   Section 2.4
        passim, 108, Figure 2
"absolute versus relational" debate   1–2, 5–6, 7,
    13–14, 52–3, 61, 67–8, 70, 149–50
a-priori principles   Chapter 5 *passim*, 8, 59–60,
    71, 83–5, Section 3.8 *passim*, Section 4.6
    passim
Aristotle   13–14, 41

Barbour, J.   152
Belot, G.   151
Ben-Menachem, Y.   97
Berkeley, G.   14, 37, 160
Bishop, R.   97
Bolzano, B.   65

Carnap, R.   9–10, 42
Cartesian physics
    absolute simultaneity in   20
    exposition   18–19
    inertia, principle of   27
    Leibniz's critique of   70

mechanical explanation, program for   18
metaphysics as foundation for physics   57
space and motion, account of   18–19, 31,
    39; Newton's critique of, and possible
    Cartesian response   30–1, 32–3, 36, 37,
    38–9
space and time substantivalism   38
substance-and-accident, theory of   37, 38
    Newton's critique of   17–20, especially
    19–20
vacuum, impossibility of   19
vortex theory of planetary motion   19, 31
Carrier, M.   80, 97
Clairaut, A.-C.   50, 53
Coffa, A.   15, 16, 97
conventionalism   Sections 3.6 and 3.7 *passim*,
    23–5, *see also* Poincaré
Copernicus, N.   19, 47
Cotes, R.   41

D'Alembert   50
Demopoulos, W.   12, 77
Descartes, R.   Section 2.3 *passim*, Section 2.6
    passim, 57
dialectic   2, 42, 45–7
Dingler, H.   80
DiSalle, R.   53, 54, 97, 151, 152

Earman, J.   7, 22, 67, 68, 96, 119–20, 151
Eddington, A. S.   15, 16, Section 4.6 *passim*
Einstein, A.   2–3, 5–6, 7, Chapter 4 *passim*
    conventionalism   9, 14–15, 102–3
    electrodynamics, on   103–5
    epistemological views   150–1
    Newton's theory, on   14–15, 35
    operationalism   109–11
    point-coincidence argument   83
    principle and constructive theories   117–20
    simultaneity, analysis of   109–11
    verificationism   101–2
Eisenstadt, J.   149

171